e-Learning Communities

Teaching and Learning with the Web

Edited by

Kwok-Wing Lai

University of Otago Press

Published by University of Otago Press
PO Box 56/Level 1 398 Cumberland Street, Dunedin, New Zealand
Fax: 64 3 479 8385
Email: university.press@otago.ac.nz

First published 2005
© Kwok-Wing Lai and individual authors 2005
ISBN 1 877372 01 3

Printed by Astra Print Ltd, Wellington, New Zealand

Errata

A font has failed to print.

p. 132   The missing word in the heading is 'Before'.

p. 198   The missing words in the heading are
          'Online' and 'Offline'.

Contents

Preface

This is a book about e-learning communities and professional development. It is written specifically for teachers and educators who are interested in using information and communication technologies (ICT) in their teaching or professional learning. I believe ICT has the potential to benefit teachers, to engage them in learning communities, and to enhance their professional practice. It is our hope that this book can provide some ideas and insights as to how ICT can be used to assist teachers for their professional development.

This book has twelve chapters. After examining some of the myths surrounding the use of ICT use in teaching and learning, I argue in the introductory chapter that one of the most promising uses it has in education is its potential to help build communities of practice for teachers. As an example, I have provided some suggestions as to how ICT might be used to support online learning communities for beginning teachers. Mark Brown in Chapter 2 challenges some of the 'taken-for-granted assumptions about the potential of e-learning in schools' and attempts to 'raise awareness of false consciousness and show how the current drive to wire New Zealand teachers with the latest ICT is highly problematic'. Chapters 3–5 focus on the professional development of teachers using the Web. Nola Campbell and Russell Yates provide a successful example of how e-learning can support a pre-service teacher education programme. Vince Ham in his chapter evaluates the Ministry of Education funded ICTPD programmes. Keryn Pratt and I in Chapter 5 discuss the importance of leadership in ICT integration. We highlight the important role of the ICT coordinator in supporting professional learning of teachers, based on a three-year study of ICT use in Otago primary and secondary schools. Chapters 6 to 9 look more closely on issues with learning communities. Reflecting on their experience in developing learning communities, Linda Selby, Ken Ryba, and Garry Falloon's chapter (Chapter 6) provides examples and insights of effective partnerships between education, business, and community. In the following chapter, Bill Anderson discusses power and control in online educational communities, highlighting issues surrounding race and gender in computer-mediated communication. Finally in this section, Judy Parr and Lorrae Ward (Chapter 8) address issues related to the building of online professional communities, based on their experience in the FarNet project. The last four chapters provide teachers with resources to enhance their understandings in using the Web in teaching and learning. Stephen Hovell (Chapter 9) examines the use of virtual field trips in the classroom, with specific reference to the LEARNZ virtual field trips. Keryn Pratt (Chapter 10) examines the ways students search the Web for information. Liz Butterfield's chapter addresses the issue of cybersafety, and finally in the last chapter Shannon Curran and Merrin Crooks-Simpson provide a guide to Web-resources on online plagiarism.

I believe in collaboration. Many people have worked closely with me in producing this volume. I am particularly grateful to the contributors who were committed to finishing their manuscripts on time. I also wish to extend my gratitude to Wendy Harrex, the managing editor of the University of Otago Press. The University of Otago has provided generous support during my tenure as Head of the Faculty of Education in the last two years, which allowed me to initiate and complete this book project. I am most grateful to my research team, particularly Dr Keryn Pratt, and Megan Anderson.

Finally, I must thank my wife, Sook-Han, for her unfailing support, love and encouragement throughout my career as a teacher and a researcher. I also wish to thank my children, Shang-Chin and Keng-Yin, for their understanding and smiles. Without their support, this book would not have been published. I am pleased to dedicate this book to them.

KWOK-WING LAI
Faculty of Education, University of Otago

Teachers using ICT: Myths and Realities

1

Kwok-Wing Lai

Introduction

We are a generation in transition. Generally reared in a print-period, but increasingly required to function electronically. We are required to teach in a way that we have never been taught (Spender, 1995).

Developing future teachers who know how to use modern leaning technologies to improve student learning is a major challenge facing our nation's teacher preparation system (Preparing Tomorrow's Teachers to Use Technology, 2002).

There has been an explosive growth of information and communication technology (ICT) use, particularly in Internet connectivity, in schools and classrooms in the last few years. For example, 92 per cent of all the classrooms in public schools in the United States (US) were connected in 2002 and the use of broadband connections in these schools increased from 81 per cent in 2000 to 95 per cent in 2002. With regard to the ratio of students to instructional computers with Internet access in public schools, it was 4.8 to 1 in 2002 (US Department of Education, 2003). The situation is similar in Britain, where the average number of pupils per computer in 2003 was 7.9 in primary schools and 5.4 in secondary schools (Office for Standards in Education [Ofsted], 2004); by July 2003, 50 per cent of all schools and 90 per cent of secondary schools were connected to broadband. In New Zealand, our own study found that 86 per cent of the primary schools in 2002 had Internet connections (Lai & Pratt, 2003), and 78 per cent of the secondary schools had Internet connections in 2001 (Lai *et al.*, 2002). These figures would be much higher now. With all these investments, there is an expectation that technology will overhaul education, serving as a panacea, or as an agent of change (even in the research community this kind of expectation prevails, e.g. in the American Research in Education Association, where there is a special interest group called Technology as an Agent of Change in Teaching and Learning).

To createe a better understanding of how teachers have been using technology in teaching and learning, in this chapter I will uncover a few myths about the use of ICT by teachers. I hope to raise a few questions, although I do not necessarily have answers to any of these questions.

Myth One:

There is no doubt that, at least in the developed world, huge resources have been invested in purchasing ICT hardware and software, and funding professional development, particularly in the last ten years. For example, in Britain in 2002–2003 alone, the funding provided by the Department of Education and Skills for ICT in schools totalled £510 million, compared with a total of £657 million from 1998–2002. In addition, a sum of £230 million was made available (1999–2003) to help increase the 'competence of all teachers in their use of ICT in teaching and learning' (Ofsted, 2004).

In the US, the situation is similar, with large funding being channelled to public schools to purchase and service technology. For example, the E-Rate programme, from 1998–2000, has provided US$5.7 billion as a universal service programme for public K-12 schools and libraries by discounting (20–90 per cent) Internet and telecommunications technologies and services (Pea, 2000). Also, during the Clinton administration, $8 billion was invested in four goals: a computer in every classroom, every classroom wired to the Internet, computer training for all teachers, and instructional software available to all students.

Even with noted increases in computer access and Internet connectivity, according to Cuban (1999), classroom use of ICT has continued to be 'uneven, slow, and of decidedly mixed variety'. He notes that in the late 1980s the vast majority of teachers were non-users, about 25 per cent were occasional users, and only 10 per cent were serious users. A decade later, in the late 1990s, after huge investments in computer hardware, only 20 per cent of secondary teachers, and about 10 per cent of primary teachers had become serious users. About 40 per cent of secondary (and 50 per cent of primary) teachers are still non-users. Becker (2000), in his TLC study, surveyed over 4000 teachers in over 1100 schools across the US and concurred that teachers' 'intractable workplace conditions do still limit widespread classroom use of computers' (p. 2). Similarly, in Britain, according to Ofsted (2004), although ICT is now 'recognised as an essential tool for learning' and there has been an increase in 'staff confidence, record resource levels and improvements in pupils' ICT capability … the spread of ICT as a tool for teaching and learning has continued at a slow, albeit steady, rate' and 'the government's aim for ICT to become embedded in the work of the schools is a reality in only a small minority of schools' (p. 6).

Unfortunately these investments have not resulted in widespread integration of technologies into the school curriculum. In our own work, in evaluating the use of technologies in New Zealand schools, we found that the lack of integration of technology into the school curriculum is rather evident. We used Knezek and Christensen's (1999) six-stage model to estimate New Zealand teachers' level of technology adoption. For primary teachers, less than 60 per cent of the teachers in 2002 were at Stages Five and Six (Stage Five being adaptation to other contexts, and Stage Six being creative application to new contexts) (Lai & Pratt, 2003). For secondary teachers, less than half (42 per cent) were at Stages Five and Six in 2001 (Lai, Pratt & Trewern, 2002).

According to Becker (2000), the most widespread professional uses of software by teachers are fairly routine: preparing handouts, writing lesson plans, and recording and calculating grades. He also reported that frequently students' experiences with school computers occur primarily in four contexts: (1) separate courses in computer education; (2) pre-occupational preparation in business and vocational education;

(3) various exploratory uses in elementary school classes; and (4) the use of word processing software for students to present work to their teachers (p. 2). He concurs with Cuban that technology has been integrated only by a 'relatively small group of academic subject-matter teachers who are significantly different than their peers in terms of teaching philosophy' (p. 18).

Why don't teachers use technologies in their teaching?

It is unfortunate that teachers are being blamed for not using technology. As pointed out by Cuban (1986) nearly twenty years ago, the use of technology in education has gone through the expectations, rhetoric, policies, and limited-use cycle. This cycle began with great expectations as to the effectiveness of the technology, coupled with predictions of positive cognitive or attitudinal outcomes. The promotion of the use of the new technology in schools was enhanced by academic studies to establish its scientific credibility. But, after a while, when the initial enthusiasm subsided and reports of limited classroom use appeared, teachers were blamed for failing to use the technology to improve learning. This cycle not only describes technology use in the pre-computer era, but also fits well with computer use in education in the last twenty to thirty years. By and large, there were four main waves of computer use in schools in the last three decades: the instructional wave (1975–85) where students were encouraged to learn from the computer as a tutor; the problem solving wave (1980–90) where the focus is shifted to students teaching the computer (e.g. Logo); the mind-tool wave (1985–95) when students learn with the computer, and the computer is used as a tool; and, finally, the media wave (1995– to present) when there is a 'convergence of media into a digital form that allows students to learn through connected computers, via the Internet' (Brown, 2004). Each of these waves has undergone a cycle similar to that described by Cuban.

Many studies have been undertaken to document why teachers do not use ICT. A long list of barriers has been suggested. We have included some in our own work (Lai, Pratt, & Trewern, 2002):

- cost of equipment
- cost of technical support
- other costs of use
- teacher knowledge of equipment
- teacher understanding about value of use
- board of trustees understanding about value of use
- availability of equipment
- availability of resource kits/learning activities
- back-up support once equipment is in use
- software/licensing
- knowledge necessary for technical planning
- knowledge necessary to integrate technology
- cost of Internet use

Even with huge investments in technology hardware, there just doesn't seem to be enough resource. The lack of access to technology is still considered by some researchers to be a major barrier to its use (Becker, 2000). To be sure, having access to a computer would be necessary if teachers were to improve their ICT skills and to integrate ICT into their teaching. In the last few years, we have seen a growth of schools and projects using laptops, both in New Zealand and overseas. For example, in New Zealand, the Ministry of Education now provides funding support for primary and secondary teachers to purchase laptop computers. Large-scale 'laptops for teachers' initiatives have also been administered in the US, Britain, and Australia (Becta, 2002a; Dean, 2001; Hart, 1999, 2002; Rockman et al., 2000; The Teacher Laptop Foundation, 2003; also see Boyd's review, 2002). It is suggested that the availability of a personal laptop computer will make a difference in teaching, as a US school district technology co-ordinator commented: 'Laptop computers allow for anytime, anywhere access needed to give the owner a sense of control over what could otherwise be a daunting device. Once the apprehension of using the computer is dispelled, real learning takes place' (Hart, 1999, ¶5).

It is important, however, to realise that providing teachers with access to hardware (including laptops) is in itself insufficient to ensure that the computers will be used. On the contrary, even with minimal computer technologies, teachers can still successfully design and implement computer-supported learning environments that are student-centred (Lai, 1993). Teachers' levels of usage and integration of technologies in their teaching are complex issues affected by an array of ICT and non-ICT factors including, among others: the attitudes and pedagogical beliefs regarding the value of the use of technology; the school and work culture; and the leadership and management of the school, as well as professional development opportunities and technical support provided by the school. Providing hardware access and professional development opportunities are important factors, but we should not expect a causal link between these two factors and a change in pedagogical approaches.

Teachers need to understand the pedagogical benefits of using technologies before they will use them in their teaching. It should be noted that technology not only provides teachers with more efficient ways of doing the 'old things' but also new ways of doing 'new things'. To fully utilise the potential of ICT in the teaching and learning processes, teachers must be willing to hold a belief system that is broadly constructivist in approach: that is, teaching and learning is an active, constructive, cumulative, and goal-oriented process (Shuell, 1986). This has been supported by a recent study conducted by Becker (2000), in which the researcher found that teachers holding a constructivist approach to learning were more willing to use technology in a deeper way.

It is also important that schools have a culture in place that encourages and supports the use of ICT (Education Review Office [ERO], 2000; Sheingold & Hadley, 1990). Our own research showed that Otago primary teachers identified a lack of vision or direction from their schools as being a barrier to their use of ICT (Lai & Pratt, 2003), and Schiff & Solmon (1999) found that in order for ICT to be effectively utilised within US schools there was a need for a well-designed technology plan driven by strong leadership. Becta (2002b) also emphasised the importance of good leadership, both in terms of ICT leadership and general school leadership, and Somekh *et al.* (2001) commented that the ICT coordinator's approach to ICT affected the approach of the whole school and, if his or her beliefs differed from that of the school's management, ICT use among teachers was generally not successful or widespread.

Our research also highlights the role of the ICT coordinator in managing changes (Lai, Trewern, & Pratt, 2002; Lai & Pratt, 2004). Although recognising the school principal as a key change agent for innovations (Fullan, 2001), the notion of shared leadership at the school level has been suggested as a better model for the implementation of innovations, with the principal not being considered the single source of direction and inspiration (Hopkins *et al.*, 1994). In our studies of twenty-five principals and twenty-five ICT coordinators in secondary schools, we have identified four key roles of the ICT coordinators: (1) planner and manager, (2) envisioner, (3) trainer, and (4) technician (see Chapter 5 in this book for a more detailed discussion on leadership issues).

Many researchers have reported that a lack of time is a key barrier to the use of ICT, commenting that teachers need time to learn how to use computers, to practise their use, to develop materials using them, to implement these in the classroom and to reflect on the implementation (Ang, 98; Capper, 1999; Cook, 1997; Glennan & Melmad, 1996; Smerdon *et al.*, 2000; National Education Association, 1999–2000). As an Organisation for Economic Co-operation and Development [OECD] report (2001) commented: 'the open applications of ICT introduce a dynamic, interactive [learning] environment that is likely to require more … planning, preparation, and one-to-one intervention than that needed for traditional curriculum delivery' (p. 74).

Teacher education
The most critical factor in the successful integration of ICTs into teacher education is the extent to which the teacher educators have the knowledge and skills for modelling the use of ICTs in their own teaching practices (UNESCO, 2001, p. 83).

It should be noted that, by and large, teacher education programmes worldwide have not been very successful in providing student teachers with knowledge, attitudes, and confidence in using technologies in classes. According to one US report (Moursund and Bielefeldt, 1999) most teacher educators do not model the use of technology in their own teaching: between 20 and 50 per cent use ICT. As well, only 20 to 50 per cent of student teachers use technology or see other teachers using it in their teaching practice. Little wonder then that in another US national survey, in 2000, only 27 per cent of the 5253 teachers interviewed felt very well prepared to integrate educational technology into the grade or subject they taught (U.S. Department of Education, 2001).

Far too often the acquisition of technical skills, rather than the pedagogy of ICT use, has been emphasised in teacher education programmes. As the Ofsted report (2004) commented, 'the need for competence with the technology drove the training rather than implications of the use of ICT for learning' (p. 8). Thus, for example, in e-learning, there is a question of whether the emphasis should be put on the 'e', or on the 'learning'. I agree with Sandholtz & Reilly (2004) that teacher education programmes should put more emphasis not on technical training but on training in how to integrate technology into the school curriculum. These authors maintain that reducing technical expectations for teachers could encourage technology use in classroom instruction. They challenge the notion that teachers have to learn the 'basics', the technical side of technology before applications: 'We don't require an automobile maintenance course before issuing a license to drive, so why do we expect teachers to master the ins and outs of personal computers before thinking about how to use them effectively in classrooms?' (p. 507).

Myth Two:

Although research has documented positive outcomes from ICT use in a variety of learning situations, most of the evidence is anecdotal, and there is currently no conclusive evidence that ICT has universal benefits or that its use is appropriate in every learning situation. Thus, in using technology in the classroom teachers should ask: 'What evidence is there that the information technology we want to use in the ways we propose to use it, for the curriculum we want to teach in the time that we have, with the teachers we employ and with the professional development and technical support we can provide will help our students learn better?' (International Society of Technology in Education, n.d.).

We need to critically question why technology should be used in the first place. Ragsdale (1982) alerted us with the law of the hammer: 'If you give a hammer to a two-year old, suddenly a lot of things will need hammering.' The hammer can become the computer, or ICT: 'If you give a computer to a teacher, suddenly a lot of things will need computing.'

There seems to be an implicit assumption that teachers should use ICT simply because it is there for them to use. Maddux (1988) called this the Everest Syndrome: 'Those who have succumbed to this syndrome believe that computers should be brought into educational settings simply because they are there. They believe further that mere exposure to computers will be beneficial to students, that computers ought to be used for any and all tasks for which software is available or imaginable, and that, if schools can only obtain a sufficient quantity of hardware and software, quality will take care of itself' (p. 5).

This view shows a lack of understanding of the importance of pedagogy as well as of the role of the teacher in the use of ICT in the teaching and learning process. ICT can be used in different ways: to increase the efficiency and the effectiveness of the 'instructional' process by transmitting knowledge, drilling skills, providing tutorials and amplifying and extending human capabilities; for student inquiry or to assist problem-solving; or to enhance creativity, to support social and cognitive development. The teacher, as facilitator, coach and manager, thus matters most. Cuban (1998) suggests that teachers should ask three questions before they use technology in their classrooms:

1. 'What do you want students and teachers to achieve in the classroom from the use of information technologies?' This question asks teachers to reflect on the goals of teaching, the nature and characteristics of ICT, and how relevant it is, or what it can do for us.
2. 'Can we reach the same goals with less cost without additional investments in technology?' This question asks teachers to consider whether technology is the best option.
3. 'What configuration of hardware, software, and Internet connections would best meet your goals and projected use of computer?' This is the 'how' question.

Can technology make a difference? Instead of being technocentric, focusing on the technology itself, I believe it is teachers' attitudes towards technology, their beliefs in teaching and learning and their teaching pedagogy that influence how students use technology and what sort of learning experience they will acquire in schools. In short, technology can be used to support a range of teaching pedagogies, from teacher-centred to student-centred, or somewhere in between. Using technology not only lets teachers and students do things faster but requires a shift of paradigms in teaching and learning. It provides an opportunity to develop a totally new relationship between the teacher and the learner. Hence the role of the teacher, rather than technology, is crucial. As pointed out in one report: 'What teachers know and can do is the most important influence on what students learn. Educators and policy makers can talk about things like governance structures, instructional methods, curricula, and standards – all crucial elements in making school more effective for children – but that connection between student and teacher is something we can get passionate about ... Teaching is what matters most' (National Commission on Teaching and America's Future, 1996).

Thus, among others, a computer-supported learning environment could include the following elements (adapted from Means & Golan, 1998):

Integration:	Curriculum integrated into existing curriculum, cross-disciplinary, problem solving, metacognitive and global.
Engaging:	Self-motivated, independent thinking, creative and self-esteem.
Real-world:	Driven by real-world issues, problems and information.
Extended:	Allow students the time necessary to study specific subject matter in depth. Quality and complexity of work produced.
Student-centred:	Allow students to choose topics, research methods, and resources, and to design their own products and presentations.
Teachers:	Facilitate student learning.
Collaborative:	Require cooperation among students on project tasks.
Multi-media based:	Products and presentations integrate multiple media, not just computing technologies.
Assessment:	Feedback to student work and final products and presentations is provided through self, peer, and teacher assessments based on curriculum standards and project goals.

Myth Three:

The use of technologies has offered many learning opportunities in the last twenty years, but teachers and researchers have paid little attention to the negative effects of ICT use. In this chapter, I raise only one concern: the health risks associated with computer use. We should note that the use of laptop computers in schools may pose an increasing risk of teachers and students developing overuse syndromes as they are expected to use a small keyboard, trackball or some pointing device for extended periods of time. In one study I have conducted, all the principals, ICT coordinators and administrative staff of both primary and intermediate schools in the Southland and Otago regions in New Zealand were surveyed (852 questionnaires posted, 362 returned). The findings showed that it was rather common for teachers, principals and administrators to experience health problems with computer use. For example, more than half the teachers reported having health problems related to the use of computers at work. The two most widespread problems were wrist pain (49 per cent) and eyestrain (44 per cent). Some examples are:

Pain/headaches—seem to relate—a dull ache—usually on left side of neck or down the back of neck. Pains in fingers—like arthritis—very sharp.

Eye strain when I first started using a computer. Rectified when I started taking a break regularly.

Back pain if not sitting in my adjustable chair properly; wrist strain if chair height not correct; eye strain if I don't adjust my blinds according to changes natural light.

A great anxiety whenever I have to use it or my inadequacies are likely to be exposed.

This last comment shows that health risks involve not only physical discomfort but psychological discomfort as well.

Teachers and principals seemed to have little awareness of health issues, as can be seen from their responses to the following questions:

- Do you watch your posture while you work on your computer?
  47 per cent answered NO
- Are you aware that the amount of lighting in your computer work area will affect you?
  35 per cent answered NO
- Do you know that there are ergonomically designed products available for computer use?
  18 per cent answered NO
- Do you know that there are software programs available to remind you the length of time you have been using the computer?
  93 per cent answered NO
- Does your school have a health and safety policy related to computer use?
  94 per cent answered NO

- Have you organised any professional development activities for your staff?
89 per cent answered NO

It is important that teachers have a good understanding of the social and health implications of ICT use in their work environment.

Myth Four:

Teachers are in some kind of dilemma when they ask their students to gather information via the Internet. On the one hand, teachers are concerned with increasing students' abilities to gather as much information via the Internet as possible. Thus, one of the most popular assignments in schools nowadays is the requirement to conduct research on the Web. There is an assumption that it is fine as long as students can retrieve some information via the Internet. Teachers seldom reflect on why students need the information in the first place or whether the information retrieved will serve some educational purpose. There is little quality assurance either. Postman (1992) maintained that it was more important for learners to pay attention to what problems could be solved by additional information, and to reflect on the implications and consequences of the process of information gathering, than simply to acquire skills to generate, retrieve, gather and distribute more information, in easier and faster ways. If we keep gathering information, without turning it into personal knowledge, soon we reach information overload, or even 'information chaos' (Postman, 1992). It has also been suggested that using the Web to conduct research may have negative effects on the development of students' critical thinking skills.

On the other hand, teachers also tend to restrict students' access to information because one of the major concerns in Internet use recently is that students may access materials that are not only unlawful to have in their possession but also may be harmful to them. I conducted a couple of studies a few years ago investigating the strategies used by teachers and parents to deal with objectionable materials on the Web. One of the most frequently used strategies was to restrict student access, by using a firewall, an intranet or some filtering programs. The use of the Web is also being tracked. For example, in one of the studies, close to three-quarters of the primary and secondary schools surveyed (196) had developed a policy to track their students' use of the Web (Lai, 2002). Teachers thus have taken a custodial role. From a constructivist learning perspective, teachers are encouraged to support students in exploring and conducting research independently, using the Internet as an information gathering tool. They are thus faced with the dilemma of how much control students should have over their information-gathering process. If students are encouraged to surf the Internet freely, they may access unscrutinised and irrelevant materials, as well as potentially objectionable or inappropriate materials. How to maintain a balance is a challenge for teachers.

It is also important to make a clear distinction between information and knowledge. Information and communication technologies can certainly enhance information retrieval, as software engineers hope to develop more intelligent search agents that are more responsive to the needs of the user; but turning information into knowledge needs the support of the teacher and a community of learners (Salomon, 2000). Salomon has outlined the difference between information and knowledge:

- information is discrete; knowledge is arranged in networks.
- information can be transmitted as is; knowledge needs to be constructed.
- information need not be contextualised; knowledge is always part of a context.
- information requires clarity; the construction of knowledge is facilitated by ambiguity, conflict and uncertainty.
- mastery of information can be demonstrated by its reproduction; mastery of knowledge is demonstrated by its novel applications.

What are the Promises?

Why do I raise all these concerns? It is not that I do not see any future in the use of ICT in education, but by raising these questions I hope teachers will reflect on their experience and become more caring and critical in their use of ICT in their teaching. There is certainly a future in the use of ICT in education. In the final section of this chapter, I wish to suggest one of the most promising uses of information and communication technologies in education: its potential for helping to build learning communities for teachers.

We all know that teaching is a rather isolating job, particularly in small and rural schools. For example, in New Zealand, nearly half (46 per cent) of all primary schools have fewer than seven teachers, and 6 per cent of the schools have only one teacher, who works alone all day without any peer interaction and support (Ministry of Education, 2002). Physical isolation very often also implies emotional and intellectual isolation. Also, teachers in general, and many beginning teachers in particular, are not well prepared to teach. For example, in a US survey, while 61 per cent of the public school teachers felt very well prepared to meet the overall demand of their teaching assignments, only 49 per cent of the beginning teachers (with up to three years teaching experience) felt they were well prepared, and only 23 per cent felt they were very well prepared to integrate educational technology in their teaching (US Department of Education, 2001). The unpreparedness and the lack of a supporting community may contribute to the high attrition rate of beginning teachers. For example, in New Zealand, it was reported that 28 per cent of the beginning teachers left their profession after three years of entry (Elvidge, 2002). In the US, it was estimated that approximately 30 per cent of new teachers leave their profession within the first five years of entry (Ingersoll, 2001).

Traditionally, formal professional development courses and workshops have been used to support teachers, but they have been criticised for being mainly short term and lacking continuity and adequate follow-up (Fullan with Stiegelbauer, 1991). They are also not particularly relevant to teachers:

'If I told you how many courses I've taken in computers, you would roll on the floor,' says

Bonnie Bracey, a teacher who was appointed by President Clinton to a federal panel on information technology from 1993 to 1995. The problem is, those courses had 'no connection to what I teach,' she adds, 'It took us a long time to figure that out' (Zehr, 1997).

Typical summer institutes for teachers do little to alter the isolated and isolating character of classroom teaching. Too often, teachers returning from these experiences have little opportunity to implement what they have learned and make significant changes in established practices in their home schools. On-going, collaborative approaches to professional development help establish a professional culture that creates self-expectations among teachers that they will be studying some aspect of practice, comparing notes on implementation, seeking new ideas, and helping each other out (Kozma & Schank, 1998).

The importance of the collaborative form of professional development within a school has been recognised (Riel & Becker, 2000). Increasingly, people recognise the importance of learning in a social context – the importance of learning together – because learning happens in the context of exchanging ideas, arguing, debating ideas and receiving feedback from other learners. Contacts with other teachers also provide the incentives to change. For example, Riel & Becker (2000) reported that teachers who had many professional contacts with other teachers at their school were three-and-a-half times more likely to employ a strong constructivist approach to learning than teachers with fewer contacts.

To collaborate with their peers, teachers need to develop and participate in their own professional communities to discuss exemplary practices and materials; to co-construct, review and publish resources that reflect new beliefs and teaching practices; and to jointly create locally relevant solutions and practices (Schlager *et al.*, 1996). In fact, teachers are already networking with their colleagues. For example, in the US, 62 per cent of the teachers surveyed in 2000 reported that they networked with teachers outside their schools, the majority a few times a year (US Department of Education, 2001).

ICT can help to create a community of practice – a professional development network – for teachers. Currently, I am in the process of developing an online learning network for beginning teachers. This online community of practice will primarily be based on a Web-based multimedia system (text, audio, video), which has the following features:

- Beginning teachers will work collaboratively with more experienced teachers (e-mentors).
- Asynchronous communication, supported by synchronous and face-to-face communication.
- A resource area where teachers can access teaching resources.
- A mentoring corner where beginning teachers can communicate with mentors in private.
- A discussion corner where beginning teachers can have synchronous and asynchronous discussion on issues by themselves.

- A coffee lounge for social interaction.
- An e-portfolio learning system where participants can work on their own, reflect on their needs and collaborate with other participants in developing teaching materials and presentations.

Making use of the capabilities of information and communication technologies, this online learning community will provide support for beginning teachers and peer collaboration by sharing exemplary practice and course materials, and also broaden beginning teachers' learning horizons.

Conclusion

When we talk about the promises of ICT use in education, we are talking about the future. It is difficult to predict the future, and our predictions are often quite wrong, as illustrated by the following two quotes:

640K ought to be enough for anyone (Bill Gates, 1981).

There is no reason why anyone would want a computer in their home (Ken Olson, president, DEC., 1977).

What I can say about technology use is that technology will change but, as teachers and educators, we should be clear about what our main objective is. Our job, as suggested by Salomon (2000), is to teach, facilitate, and support students to become 'independent, mindful thinkers, skilled in life long learning, capable of intelligently handling complex problems alone and in teams, and guided by some social values that transcend egotistic benefits'. Technology should be used to support teachers to achieve this.

Acknowledgement

The author wishes to acknowledge the editorial support of Megan Anderson and Dr Keryn Pratt. An earlier draft of this chapter was presented as a keynote address at the EduCATE 2004 international conference, Kuching, Malaysia, August 2004.

References

Ang, C. (1998/9). *Evaluation report: The FY98 Professional Development Days (PDD) Program.* Available at http://www.palmbeach.k12.fl.us/9045/pdd.htm

Becker, H.J. (2000). *Findings from the teaching, learning and computing survey: Is Larry Cuban right?* Available at http://www.crito.uci.edu/tlc/findings/ccsso.pdf

Becta (2002a). *Laptops for Teachers Initiative: Guidance for the Laptops for Teachers (LfT) Initiative for the 2002–2003 Financial Year.* Coventry: Becta.

Becta (2002b). *Primary Schools – ICT and standards: A report to the DfES on Becta's analysis of national data from OFSTED and QCA.* Available at: http://www.becta.org.uk/research/reports/ictresources.html

Boyd, S. (2002). *Literature review for the evaluation of the Digital Opportunities projects: Report to the Ministry of Education.* Wellington: New Zealand Council for Educational Research.

Brown, M.E. (2004). The study of wired schools: A study of Internet-using teachers. A thesis submitted as fulfillment of the requirements for the degree of Doctor of Philosophy. Palmerston North: Massey University.

Capper, P. (1999). *Information technology purchasing strategies in schools: Scoping report.* Wellington: Ministry of Education.

Cook, C.J. (1997). *Critical issue: Finding time for professional development.* Available at: http://www.ncrel.org/sdrs/areas/issues/educatrs/profdevl/pd300.htm

Cuban, L. (1986). *Teachers and machines: The classroom use of technology since 1920.* New York: Teachers' College Press.

Cuban, L. (1998). *The pros and cons of technology in the classroom. Part 2: Cuban Speech.* Available at: http://www.tappedin.org/info/teachers/debate.html

Cuban, L. (1999). *Why are most teachers infrequent and restrained users of computers?* Report from the Fifth Annual Public Education Conference, Vancouver. Available at: http://www.bctf.bc.ca/parents/PublicEdConf/report99/appendix1.html

Dean, K. (2001). *Smart idea: Laptop for teachers.* Available at: http://wired.com/news/school/0,1383,46987,00.html

Education Review Office [ERO] (2000). *In-service training for teachers in New Zealand schools.* Available at: http://www.ero.govt.nz/Publications/pubs2000/InServiceTraining.htm

Elvidge, C. (2002). Teacher supply. Beginning teacher characteristics and mobility. Unpublished paper for Ministry of Education.

Fullan, M. (2001) *The new meaning of educational change.* New York: Teachers College Press.

Fullan, M. with S. Stiegelbauer (1991). *The new meaning of educational change.* London: Cassell.

Glennan, T.K. & Melmad, A. (1996). *Fostering the use of educational technology: Elements of a national strategy.* Santa Monica, CA: RAND Publications.

Hart, P. (1999). *Teacher Laptop Program: Integrating Technology into the Curriculum.* Available at: http://www.edc.org/LNT/news/Issue8/field.htm

Hopkins, D., Ainscow, M., & West, M. (1994). *School improvement in an era of change.* London: Cassell.

Ingersoll, R. M. (2001). *Teacher turnover, teacher shortages and the organization of schools.* Washington: University of Washington, Center for the Study of Teaching and Policy.

Knezek, G. & Christensen, R. (1999). Stages of adoption for technology in education. *Computers in New Zealand Schools*, 11(3), 25–28.

Kozma, R. & Schank, P. (1998). Connecting with the twenty-first century: Technology in support of educational reform. In D. Palumbo and C. Dede (eds), *Association for Supervision and Curriculum Development 1998 Yearbook: Learning and technology*, Alexandria, VA: ASCD, pp. 3-27.

Lai, K.W. (1993). Minimal computer technologies and learner-centred environments: Some New Zealand experience. *Computers & Education*, 20(4), 291-7.

Lai, K.W. & Pratt, K. (2003). *Learning with Technology II: Evaluation of the Otago Primary Schools Technology Project.* Dunedin: The Community of Otago Trust.

Lai, K.W. & Pratt, K. (2004). ICT leadership in secondary schools: The role of the computer coordinator. *British Journal of Educational Technology*, 35(4), 461–75.

Lai, K.W., Pratt, K. & Trewern, A. (2002). *Use of information and communication technology in Otago secondary schools: Phase 1 of a three-year study.* Dunedin: Community Trust of Otago.

Lai, K.W., Trewern, A. & Pratt, K. (2002). Computer coordinators as change agents: Some New Zealand observations. *Journal of Technology and Teacher Education*, 10(4), 539–51.

Maddux, C. (1988). Preface to a special edition on assessing the impact of computer-based instruction. *Computers in the Schools*, 5(3/4), 1–10.

Means, B. & Golan, S. (1998). *Transforming teaching and learning with multimedia technology.* Menlo Park, CA: SRI International.

Ministry of Education (2002). *Digital horizons: Learning through ICT.* Wellington: Ministry of Education.

Ministry of Education (2002). *Preparing Tomorrow's Teachers to Use Technology.* Available at: http://www.ed.gov/teachtech/

Moursund, D. & Bielefeldt, T. (1999). *Will new teachers be prepared to teach in a digital age? A national survey on information technology in teacher education.* Santa Monica: Milken Exchange on Education Technology.

National Commission on Teaching & America's Future (1996). *What matters most: Teaching for America's future.* New York. Available at http://www.nctaf.org

National Education Association (2000). *It's about time.* Available at: http://www.nea.org/bt/3-school/time.pdf.

Office for Standards in Education [OFSTED] (2004). *ICT in schools: The impact of government initiatives five years on.* Available at: http://www.ofsted.gov.uk/publications/index.cfm?fuseaction=pubs.summary&id=3652

Organisation for Economic Co-operation and Development [OECD] (2001). *Learning to change: ICT in schools.* Paris: OECD.

Pea, R.D. (2000). *Bridging the digital divide: Technology, equity, and K-12 learning.* Available at: http://www.ccst.us/ccst/pubs/cpa/bdd/BDDreport/BDD09.html

Postman, N. (1992). *Technology: The surrender of culture to technology.* New York: Alfred A. Knopf.

Ragsdale, R. (1982). *Computers in the schools.* Toronto: OISE.

Riel, M. & Becker, H. (2000). *The beliefs, practices, and computer use of teacher leaders.* Paper presented at Annual Meeting of the American Educational Research Association, New Orleans, 2000.

Rockman et al. (2000). *A more complex picture: Laptop use and impact in the context of changing home and school access.* Available at: http://rockman.com/projects/laptop/laptop3exec.htm#top

Salomon, G. (2000). *It's not the tool, but the educational rationale that counts.* Keynote address at the 2000 Ed-Media Meeting, Montreal. Available at: http://www.constuct.haifa.ac.il/~gsalomon/edMedia2000.html

Sandholtz, J.H. & Reilly, B. (2004). Teachers, not technicians: Rethinking technical expectations for teachers. *Teachers College Record,* 106(3), 487–512.

Schiff, T.W. & Solmon, L. C. (1999). *California Digital High School Process Evaluation (Year One Report).* Santa Monica: Milken Exchange on Education Technology.

Sheingold, D. & Hadley, M. (1990). *Accomplished teachers: Integrating computers into classroom practice.* New York: Bank Street College of Education.

Shuell, T. (1986). Cognitive conceptions of learning. *Review of Educational Research,* 54, 411–36.

Smerdon, B., Cronen, S., Lanahan, L. *et al.,* (2000). *Teachers' tools for the 21st century* (No. NCES 2000–102). Washington, DC: National Center for Education Statistics, US Department of Education.

Somekh, B., Barnes, S., Triggs, P. *et al.* (2001). *NGfL Pathfinders. Preliminary report on the roll-out of the NGfL programme in ten pathfinder LEAs* (no. 2). Nottinghamshire: Department for Education and Skills.

Spender, D. (1995). *Nattering on the net: Women, power and cyberspace.* Melbourne: Spinifex Press.

The Teacher Laptop Foundation. (2003). *The laptop technology gift.* Available at: http://www.teacherlaptop.org/laptop.html

UNESCO (2001). *Winds of change in the teaching profession.* Paris: French National Commission for UNESCO.

US Department of Education, National Center for Educational Statistics (2003). *Internet access in US public schools and classrooms: 1994–2002.* Available at: http://necs.ed.gov/pubsearch

US Department of Education, National Center for Educational Statistics (2001). *Teacher preparation and professional development: 2000.* Available at: http://necs.ed.gov/pubsearch

Zehr, M.A. (1997). *Teaching the teachers.* Available at: http://www.edweek.org/sreports/tc/teach/ten.htm

Telling Tales Out of School: The Political Nature of the Digital Landscape

2

Mark Brown

The intention of this chapter is to challenge some of the assumptions about the potential of e-learning in schools. In this context, the term 'e-learning' refers to the latest wave of policy initiatives that promote the adoption and implementation of new educational technologies for learning and teaching purposes. By telling tales out of school, the chapter attempts to raise awareness of false consciousness and show how the current drive to wire New Zealand teachers with the latest ICT is highly problematic. It is far more problematic than is evident in most professional magazines, educational conferences and Ministry of Education policy briefs, which typically celebrate the benefits of new computer technology in schools.

The chapter is in three parts. In the first section, the growth of e-learning is located in the backdrop of the wider debate surrounding the role of ICT in education. This debate must be taken seriously in the face of several high-profile attacks on the use of computers in schools. In the second part, the discussion illustrates how the ICT-related school reform movement has been dominated by celebratory discourses. The overselling of ICT in New Zealand has been at the expense of deeper intellectual debate over the way in which new computer technology may affect teachers' lives and work culture – for better and worse. In the final section, the chapter goes beyond the current orthodoxy of optimism (Selwyn & Gorard, 2002) by locating the e-learning phenomenon within the deeply contested and inherently political nature of the curriculum itself. A conceptual framework which helps teachers and teacher educators read the discourses of persuasion behind pressures to boost capacity, increase bandwidth and catch the knowledge wave, is proposed.

My main objective is to locate e-learning within the bigger picture of educational reform. By peeling away the competing and coexisting discourses of the ICT-related school reform movement, the discussion makes the case for greater pedagogical activism and underscores the importance of adopting critique as a permanent philosophical ethos. In this regard, it encourages further critical dialogue around a number of searching questions:

- Who is telling the e-learning story and why?
- How are they telling the e-learning story?
- What is it they are telling/promoting about e-learning?
- How have different people responded to the e-learning message?
- What is missing? Whose voice is not being heard? Whose story is not being told?

Critiques of the ICT movement need to go beyond a simple dichotomy of illusory hype versus pessimistic Armageddon (Abbott, 2001). Put simply, e-learning is neither demon nor panacea, as such binary positioning underestimates the complexity of the digital landscape. On the one hand, the chapter is highly critical of the hidden curriculum behind the growth of e-learning in schools. Having said that, concerns raised about the ICT movement should not be construed as further ammunition for a neo-conservative backlash. The position advanced is not neo-conservative. In a similar vein, there is a danger that concerns about the growth of e-learning will be appropriated to feed a new moral panic. It is not my intention to lend support to some of the unsubstantiated fears about new technological developments. On the other hand, the chapter attempts to avoid the pedagogy of the depressed by promoting the language of possibility – albeit from a more critical perspective. In this way, the critical views expressed go well beyond any latent mistrust of new computer technology; they cannot easily be dismissed by the advocates of ICT as nothing more than a neo-Luddite response to technological progress.

Overall, this chapter offers a more realistic perspective not so skewed by the hyperbole associated with the potential of e-learning in schools. This perspective is informed by a type of critical realism, which places more attention on the why of e-learning rather than the how of translating the rhetoric into practice, and adopts the view that more talk and critique is required before blind faith in the potential of the ICT movement steers the teaching profession further away from the real goals of education – that is, promoting equality, fairness and social justice.

The Technology Debate

There is no doubt that ICT is one of the most spectacular technological developments of the twentieth century. A new digital revolution is underway that hi-tech proponents proclaim is poised to transform our classrooms. As Bill Gates (1995) once pronounced: 'We stand at the brink of another revolution. This one will involve unprecedentedly inexpensive communication; all the computers will join together to communicate with us and for us. Interconnected globally, they will form a network, which is being called the information highway' (pp. 3–4).

In all of its manifestations, the information highway is part of a new epoch of human civilisation. It has huge implications for schools and enormous potential as a pedagogical innovation. This is without dispute. However, this revolution is far more problematic than is typically acknowledged by the proponents of the so-called digital age. In Postman's (1993) terms, 'every technology is both a burden and a blessing; not either-or, but this-and-that' (p. 5). In a similar vein, Rosen (1998) reminds us that 'every great transformation leaves social debris in its wake' (p. 37). Thus, it is not surprising that there has been a steady rise in the number of people and publications beginning to question the wisdom of the substantial investment in new computer technology. Armstrong & Casement (1998) believe, for example, that it is scandalous so much money has been allocated for computers and Internet access with so little serious evaluation. As far as they are concerned: 'A generation of children have [*sic*] become the unwitting participants in what can only be described as a huge social experiment' (Armstrong & Casement, 1998, p. 2).

They go on to observe that our insatiable appetite for new computer technology is such that one would think nothing else worthwhile is happening in schools. Although no empirical evidence is offered to support this conclusion, Armstrong & Casement believe some basic questions about the educational value of computers remain unanswered. In their words, we suffer from 'illusions of progress'.

In 1997, the level of public concern over the ICT movement was originally heightened when the *Atlantic Monthly* attacked the spurious evidence supporting the 'computer delusion' in schools (Oppenheimer, 1997). After a thorough investigation of the literature, Oppenheimer (1997) concluded: 'There is no good evidence that most uses of computers significantly improve teaching and learning...' (p. 45).

Although there is a grain of truth in this conclusion, it exaggerates the evidence in the opposite direction (Reeves, 1998). Such a blanket statement gives insufficient attention to the instructional context, as the computer is not a monolithic machine that teachers use in a uniform manner. Put bluntly, most teachers and researchers know it is technocentric to think that ICT alone can significantly improve learning. This is a vital point Oppenheimer fails to acknowledge in his critique of the literature.

What he also fails to acknowledge is 'that such pedagogical enhancements would often be impossible without the capabilities of new technology' (Reeves, 1998, p. 52). Therefore, this infamous attack on the ICT-related school reform movement contains some serious flaws. Despite this, the computer delusion article helped to fuel a growing neo-conservative backlash against the use of computers in schools, which has gained renewed momentum since publication of *The Flickering Mind: The false promise of technology in the classroom and how learning can be saved* (Oppenheimer, 2003).

Although the tendency is to dismiss these latest attacks as uninformed and poorly researched analyses of the ICT movement, such publications contribute greatly to further critical analysis. *The Flickering Mind* is a timely reminder of the need for teachers to continually question and justify the faith they place in new computer technology. In this regard, attacks on the use of computers in schools offer teachers a rich source of critical reflection. They should not be dismissed out of hand, as Oppenheimer (2003) helps to spotlight the fragility of the pedagogical rationale and the serious flaws of the social, economic and vocational rationales, which together combine to form the language of persuasion championing the educational use of technology.

In terms of the pedagogical rationale, Oppenheimer, along with Cuban (2001), Healy (1998), and so forth, has raised serious questions about the overselling of new computer technology. The so-called critics have brought attention to an alternative body of literature claiming that computer use may be detrimental to our brains, bodies and spirits. This type of analysis, which lacks solid research evidence, is supported by Stoll (1999), who argues that computers send the wrong message by making learning appear colourful and fun when it actually requires hard work and discipline. On the surface, this observation may resonate well with some parents and teachers. However, there is an element of a new moral panic embedded in this recall to the protestant work ethic.

In a similar vein, nevertheless, Cordes & Miller (2000) claim in their controversial 'Fool's Gold' report on the use of computers in early childhood education that 'The computer – like the TV – can be a mesmerizing babysitter' (p.3). They go on to say that:

'Those who place their faith in technology to solve the problems of education should look more deeply into the needs of children. The renewal of education requires personal attention to students from good teachers and active parents, strongly supported by their communities. It requires commitment to developmentally appropriate education and attention to the full range of children's real low-tech needs – physical, emotional, and social, as well as cognitive' (p. 4).

Once again, there is an element of truth in this reactionary response to the dangers of new computer technology, but what Cordes & Miller fail to acknowledge is that teachers can use ICT to enhance the holistic education of their students (Abbott, *et al.*, 2001). Thus, the 'Fool's Gold' report, like much of the earlier research on the effects of television, ignores the context of computer use. Ironically, all these so-called critics are guilty of assigning too much attention to the technology itself, which is exactly what they accuse the proponents of the digital revolution of doing.

In spite of this criticism, on another front Oppenheimer (1997) reminds us that it is extremely shortsighted to focus on today's idea of what tomorrow's jobs will be. Stoll (1999) takes the critique of the vocational rationale – a growing proportion of the workforce will require computer skills – one step further by illustrating how the adoption of new computer technology has resulted in the de-skilling of many jobs. Arguably, most so-called 'hi-tech jobs' involve little more than passing a tin of baked beans over a bar-code scanner in the supermarket (Stoll, 1999).

Far from being skilled technicians, Armstrong & Casement (1998) claim that the vast majority of computer operators are nothing more than typists doing mundane repetitive work. This line of argument requires further empirical analysis, but it suggests that the information highway has created the demand for a large technical class that is highly trained to do 'mind-numbingly boring' work (Roberts, 1998, cited in Healy, 1998). Of course, it just so happens that this type of work is reasonably well paid for in comparison with many traditional blue-collar jobs.

Although the Internet is a powerful icon of the new digital economy, Kirkpatrick & Cuban (1998) question whether the ICT movement will help create the type of critically informed students and citizens we seek. They point out that schools are not simply agents of social and cultural reproduction where future workers learn how to earn. Put another way, the digital curriculum may be preparing students to make a living, rather than educating young minds to make a life that will contribute to creating a better society (Postman 1996).

In a powerful analogy, Postman (1996) draws a parallel between the computer and the invention of the motor vehicle: 'What we needed to know about cars – as we need to know about computers, television, and other important technologies – is not how to use them but how they use us' (p. 44). According to Postman, what we really needed to think about when motor vehicles were first invented was not how to drive them but what they would potentially do to '…our air, our landscape, our social relations, our family life, and our cities' (p. 44). This analogy strikes at the heart of the debate surrounding the uncritical adoption of new computer technology in schools. It shows that the car is not simply an internal combustion engine with seats in a steel casing on wheels (Henwood *et al.*, 2000). Indeed, whether people own cars at all – and if they do, their ages, makes, and colours – provides meaning for them and others about who they are and what they value. The lesson is that ICT is not neutral; it is an inherently value-laden cultural artefact.

Despite this, the metaphor of the computer as a neutral learning tool is dominant throughout the teaching profession. As Moss (2002) writes in a recent New Zealand Education Institute (NZEI) publication: 'ICT is just another tool. You choose the best tools for the job – it might be the telephone, the Internet or a library book. If you get into that frame of thinking and your students have that frame of thinking, it is much easier to integrate ICT across the curriculum' (p. 3).

This conception of ICT reflects a form of social or cultural determinism in which the way the tool is used is far more important than the tool itself. There is no conception of the tool having an effect, either good and bad, over and above how teachers use it. As Burbules & Callister (2000) write:

> Tools do not only help us accomplish (given) purposes; they may create new purposes, new ends, that were never considered before the tools made them possible. In these and other ways tools change the user: sometimes quite concretely, as when the shape of stone tools became a factor in the evolution of the human hand ... Tools may have certain intended uses and purposes, but they frequently acquire new, unexpected uses and have new, unexpected effects. What this suggests is that we never simply use tools, without the tools also 'using' us (p. 6).

It follows that the popular metaphor of technology as progress has been challenged seriously in the context of the above debate. Those critics who adopt an extreme position warn that the computer has become the new god; it has all the features of a dangerous cult (Postman, 1996). Roszak (1994) first drew this parallel when he wrote: 'Like all cults, this one has the intention of enlisting mindless allegiance and acquiescence. People who have no clear idea of what they mean by information, or why they should want so much of it, are nonetheless prepared to believe that we live in the Information Age, which makes every computer around us what the relics of the True Cross were in the Age of Faith: emblems of salvation' (p. x).

In sum, there is considerable debate in the international literature over the rise and rise of ICT in schools. There are some serious concerns and well-articulated arguments both for and against the ICT-related school reform movement. In the backdrop of this debate, irrespective of one's position, the growth of ICT is contestable and teachers must treat the latest e-learning wave as problematic.

Lack of New Zealand Debate

This section shifts the technology debate to the New Zealand context. In the context of the debate described above, it explores two questions. First, how is the contested and problematic nature of ICT reflected in the New Zealand policy discourse? Second, how is the technology debate reflected in the New Zealand professional discourse? More straightforwardly, how does this debate manifest itself within the ICT teacher education community in New Zealand?

Policy discour

A closer look a. *Digital Horizons*, the Ministry of Education's ICT strategy for Schools, reveals the dominance of the tool metaphor. There Trevor Mallard, Minister of Education writes: 'The Government has been quick to seize on the importance and practical benefits of digital technology as a key tool for 21st century teaching and learning' (p. 2).

In keeping with the pragmatism of 'third way' politics, there is no acknowledgment of the non-neutrality of new computer technology. The potential negative and unanticipated effects of ICT receive no consideration. At the national level, therefore, policymakers have actively promoted the metaphor of computer as tool as unproblematic. Following on from this, it is highly misleading for policymakers to state in the 2002–04 iteration of the ICT Strategy that: 'The expansion of ICT is driving significant changes in many aspects of endeavour throughout the world' (Ministry of Education, 2002, p. 6).

Such technological determinism conveys a sense of inevitability that as technology changes so society follows. Although the technology is having a dramatic effect, this type of statement ignores the powerful forces that are at least partly behind the drive to equip students and workers with new types of digital literacy. The growth of ICT is not on an independent trajectory (Clegg, *et al.* 2003), as it is intertwined deeply with the globalisation movement, the rise of neo-liberalism, the celebration of technology consumption and ecologically destructive cultural patterns (Bowers, 2000). Put bluntly, the expansion of ICT in the context of these global forces is potentially brutal and socially destructive.

There is no reference, nevertheless, to these concerns in the Ministry's ICT Strategy. Instead, Trevor Mallard states: 'My vision for education is: for all students, irrespective of their backgrounds, to develop the knowledge, understandings, skills, and attitudes to participate fully in society, to achieve in a global economy, and to have a strong sense of identity and culture' (Ministry of Education, 2001, p. 4). Although this vision acknowledges the presence of competing local, national and global interests, there is a fundamental tension between retaining a strong sense of identity and culture on the one hand, and full participation and achievement on an international scale in a global economy on the other. In short, the retention of a sense of identity and culture is increasingly problematic in the face of globalisation (Hlynka, 2003).

While the above vision was revised when *Digital Horizons* was published, it promotes the goal of full *participation* rather than critical education *for* citizenship. This suggests that we should settle for citizens who will uncritically participate in the type of global society we have today. There is a key difference between passive and active citizenship, and education *for* citizenship as opposed to education *about* and *through* citizenship (Selywn, 2002). In my view, teachers need to use ICT to promote education *for* critical citizenship, in which people can contribute to building a socially just society and very different type of global community than the troubled one that exists today.

Yet, the policy discourse continues to treat ICT as unproblematic. In the latest Schooling Strategy Discussion Document teachers are told that: 'To "future-proof" schooling, the government is currently committed to ... continuing to support teachers, school leaders and boards of trustees to realise the learning opportunities presented by new technologies, through the ICT Strategy' (Ministry of Education, 2004a, p. 25).

One could equally argue that the concept of future-proofing in the context of ICT and the global economy is an oxymoron. After all, you have only to look at the relatively short life of the ICT strategy. The problem is that policymakers are presenting the adoption of ICT as one of the solutions to future-proofing despite technology consumption being a major barrier to long-term sustainability. Notably, the concept of sustainable education is conspicuous by its absence in the latest schooling strategy. Arguably, the more people utilise new computer technology and the more enmeshed New Zealand becomes in the global economy, the more dependent we become on changes to the technology and the more vulnerable New Zealand becomes to global forces and threats.

This returns us to the issue of globalisation. The reason so many people are opposed to globalisation is that a predictable outcome is the loss of nation-state sovereignty, the erosion of local autonomy and, in turn, a weakening of the definition of the 'citizen' as a unifying concept characterised by precise roles, rights and obligations (Codd, 2002). In economic terms, ICT is the digital lubricant of the globalisation movement, which explains why the OECD is such a major force in promoting the adoption of ICT in schools. Although globalisation is not all bad (Brown, 2003) in educational terms, the neo-liberal forces driving this movement are reflected in an agenda that advances particular policies for evaluation, financing, assessment, standards, teacher education, and so forth (Codd, 2002).

In sum, the lesson from the policy discourse is that teachers need to start thinking about the way in which ICT is a political and ideological vehicle for shaping a wider social, economic and educational agenda. The latest e-learning wave cannot be separated from the drive to create a new knowledge economy, because the pedagogical rationale promoting the use of new computer technology in schools is infected by the language of a kind of 'enterprise constructivism' – that is, the celebration of individualism, entrepreneurship and learning for the real (unjust) world.

Professional discourse

We now shift our attention to the professional discourse. This section draws on an example from the 2004 series of education conferences hosted by the Telecommunications Users Association of New Zealand (TUANZ). What is TUANZ? To quote from its website: 'Formed in 1986, TUANZ is the non-profit organisation that represents business telecommunications interests and is committed to ensuring that New Zealand develops a competitive and innovative telecommunications market. Our Association has more than 500 corporate members, who represent a cross-section of the major business users of telecommunications' (TUANZ, 2004).

In a nutshell, TUANZ (2004) describes itself as 'The User's Champion in the Knowledge Economy'. From this perspective, it is not surprising that TUANZ would present a one-sided message on the benefits of ICT in education. Indeed, its involvement in education underscores the above point that the latest e-learning wave must be understood from a political-economy perspective. A discourse analysis of the titles and abstracts at each of the TUANZ conferences reveals absolutely no acknowledgment of the problematic nature of ICT in schools. In short, the heated technology debate is non-existent within the TUANZ movement.

Taking this analysis a step further, over a decade ago, McMillan (1993) wrote a controversial article in which he claimed that men are typically over-represented at New Zealand educational computing conferences. He based this observation on the gender of participants at the New Zealand Computers in Education Society Conference, Nelson. Although counting the gender of participants is a crude measure of whose voice is shaping the ICT movement, this type of analysis reveals an interesting pattern.

Figure 1 helps to establish the demographic profile of the teaching profession. The most recent data from the Ministry of Education (2003) shows that at the primary level 14 per cent of classroom teachers are men. This figure increases to 19 per cent when inclusive of principals and senior management. At the secondary level, the gender profile is distributed more evenly with men making up 39 per cent of classroom teachers. Overall, the key point is that more than 70 per cent of New Zealand teachers are women.

In contrast, Table 1 presents the percentage of male and female presenters at the different TUANZ conference venues. These figures are taken from the 2004 conference programme listed on the TUANZ website. Across the board, men are over-represented at each conference, both numerically and proportionately in relation to the wider profile of the teaching profession. Overall, 61 per cent of the presenters are men. Of course, this may be just an atypical example – albeit the results are consistent throughout the country.

Figure 1: Gender profile of New Zealand teachers at March 2002

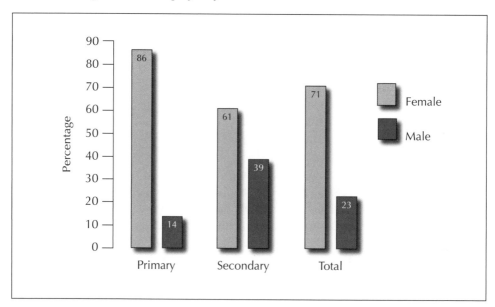

Table 1: Percentage of male and female presenters at
TUANZ 2004 Education Conference series

Location	Female	Male
Auckland – Primary	46% (n=13)	54% (n=15)
Auckland – Secondary	33% (n=10)	67% (n=20)
Hamilton	42% (n=11)	58% (n=15)
New Plymouth	32% (n=8)	68% (n=17)
Napier	41% (n=11)	59% (n=16)
Nelson	46% (n=13)	54% (n=15)
Greymouth	36% (n=11)	64% (n=20)
Total	39% (n=77)	61% (n=118)

In a similar vein, Ham and Wenmoth (2002) report that male teachers make significantly greater use of Te Kete Ipurangi (TKI). Is this a reflection of the disproportionate number of men in senior management positions for whom Internet access and conference attendance is a regular feature of their jobs? Alternatively, does it reflect the relative dominance of male ICT coordinators in New Zealand schools? In their recent evaluation of the *Otago Technology Project,* for example, Lai & Pratt (2003) found that men hold 40 per cent of the ICT coordinator positions in primary schools. Notably, at the secondary level, 78 per cent of the ICT coordinators were male (Lai, Trewern & Pratt, 2002). These findings might say something about the nature of ICT itself, which is claimed to privilege a masculine gaze (Yelland & Rubin, 2002). Although we need to be wary of deficit thinking, as experience may be far more important than gender alone, these data remind us of the way in which ICT is part of a much wider social practice.

Returning to the lack of overall debate in the professional discourse, an analysis of the titles and abstracts at this year's Learning@schools conference hosted by the Ministry of Education (2004b) in Rotorua, once again highlights the one-sided nature of the programme. Based on the information available, there was no obvious attempt to engage the new ICT professional development (ICTPD) clusters in genuine dialogue over the problematic nature of e-learning in the context of the wider technology debate. Generally speaking, the use of ICT in schools was taken for granted and seen as a good thing. As the conference website states: 'The conference programme will reflect much of the best that is currently happening in ICT in New Zealand ...' (Ministry of Education, 2004b).

Although the above evidence offers only a small slice of the professional discourse and it may underestimate the level of critique by individual teachers, the key point is that educational conferences typically present a selective version of what is being said about the ICT-related school reform movement. On the whole, these types of conferences are conceptualised from an unproblematic 'computer as tool' and 'technology as progress' perspective. There is little or no consideration of the potential negative or unanticipated distal effects of ICT on teachers' lives and work culture. For instance, there is little evidence to suggest that teachers are being invited to discuss issues such as increased workloads and the intensification of work due to new computer technology. Moreover, the potential displacement costs from focusing on ICTPD at the expense of other curriculum activities and pedagogical innovations is rarely the subject of discussion. Rather than problematising the ICT movement, the emphasis has been on closing the gap between the already 'connected' teachers and their 'disconnected' colleagues, which reflects a deficit model of teacher education.

In sum, to answer the original question – how is the problematic nature of ICT reflected in the New Zealand policy and professional discourse – by and large it is not. The overriding impression is that teachers should be embracing new computer technology and using it in so-called constructivist ways rather than adopting critique as a permanent philosophical ethos. As a result, misinformation, dissembling language and even propaganda are preventing them from understanding the hidden curriculum and non-educational intention of the ICT-related school reform movement.

Peeling Away the Orthodoxy of Optimism

This last section goes beyond the orthodoxy of optimism by locating ICT within the deeply contested nature of the curriculum itself. As Apple (1992; cited in Peters & Marshall, 2004, p. 111) reminds us: 'Education is deeply implicated in the politics of culture. The curriculum is never a neutral assemblage of knowledge, somehow appearing in the texts of classrooms of a nation. It is always part of a selective tradition, someone's selection, some group's vision of legitimate knowledge' (p. 34).

Thus, the question is: Whose vision of legitimate knowledge is behind the growth of ICT in schools? This is a difficult question to answer because the dominant groups behind the ICT movement are seeking to advance their own agendas not through brute force but through persuasion. In recent years, this has involved persuading teachers, or at least a reasonable proportion of the teaching profession, that the investment in ICT makes good sense. A core group of teachers has been recruited, often unwittingly, as the foot soldiers of the digital revolution since they are likely to have far more influence on their colleagues than those promoting the non-educational agenda.

Thus, the concept of hegemony – in which dominant groups in society seek to establish common sense meanings, define what counts as legitimate areas of agreement and disagreement, and shape the political agendas made public and discussed as possible (Apple, 2003) – is central to peeling away and mining through the deeper layers of the digital landscape. When the terrain is exposed from this critical perspective, the use of ICT is rooted in several competing and coexisting discourses with conflicting ideological and pedagogical assumptions. The main discourses can be described as: (1) reproduction; (2) reschooling; (3) deschooling; (4) reconceptualist; and (5) socio-cognitive (Codd *et al.*, 2002). Figure 2 shows how these discourses occupy different spaces of an hourglass in which the digital sand currently flows towards the direction of the so-called knowledge economy, as opposed to the knowledge society, through the reproduction and reschooling discourses.

Figure 2: The competing and coexisting discourses

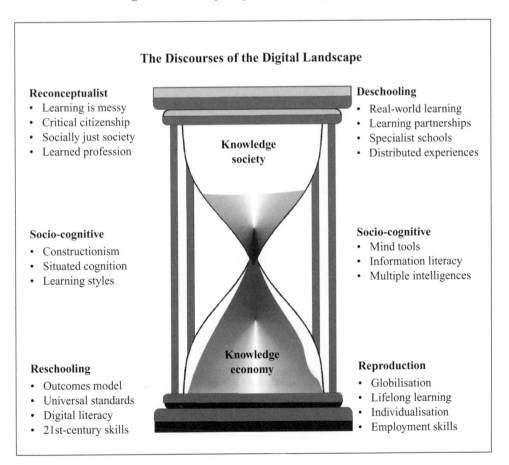

The reproduction discourse assumes that schools are still the major agent of social and cultural reproduction. Accordingly, this discourse places strong emphasis upon the preparation of students with skill and knowledge required for future employment. The current response to the shortage of ICT skills can be understood as a crude attempt to prepare a pool of future 'hi-tech' employees to keep business competitive in the knowledge economy. In this discourse, schools are the producers of human capital needed by the economy in the form of a trained workforce. It follows that: 'Teachers are then readily seen as the specialized workforce producing the larger workforce' (Connell, 1995; cited in Smyth *et al.,* 2000, p. 6).

The reproduction discourse is pervasive throughout government policy, as illustrated in the following quote: 'Mr Mallard says the use of ICT to support student learning is a major focus for Government, as well as an important means of developing a more innovative economy' (Principals Today, 2004, p. 28).

The fundamental purpose of education in a knowledge society is not considered. Put simply, ICT is promoted as the digital lubricant for a new kind of fast capitalism. The discourse is constructed around an economic and social reproduction rationale, albeit cleverly disguised in educational terms, rather than truly promoting the goal of creating critical thinkers, critical consumers and critical citizens.

The reschooling discourse promotes an outcomes model of education with an interesting alliance between the neo-conservatives and the neo-liberals. This discourse represents a shift from controlling the inputs – that is, teacher qualifications, curriculum content and instructional methods – to standardising the outputs. The discourse seeks to reschool the traditional curriculum by advancing universally high standards through the language of excellence, modernisation and technology as progress. Standards are seen as the solution to reducing the current achievement gaps, which ironically ICT makes possible through more efficient reporting of student progress.

On another front, the move to develop standardised learning objects that can be shared across national boundaries is a potentially dangerous form of teacher-proofing, which may lead to a more internationalised curriculum. Arguably, such a curriculum helps to promote a more mobile workforce capable of shifting from one market to another, and thereby reducing temporary labour shortages. Thus, the reschooling discourse is about preparing a sufficient pool of hi-tech workers for the low-tech electronic factories of the future. In the words of President Bill Clinton: 'Frankly, all the computers and software and Internet connections in the world won't do much good if young people don't understand that access to new technology means … access to the new economy' (cited in Cuban, 2001, p. 18).

The deschooling discourse contains an interesting mix of future-focused initiatives that give the impression schools are losing their monopoly on the enterprise of learning. This discourse advances the view that students live in a new digital age where e-learning is chipping away at the walls of the classroom.

The language of deschooling is behind moves to establish digitally enhanced classrooms, homework and community learning centres, and so forth. Although these initiatives recognise that learning can occur in non-traditional settings, they often treat education as a purchasable commodity and few advance a genuine deschooling agenda (Codd *et al.*, 2002). This point is illustrated in the uptake of laptop computers, where many of these initiatives promote new ways of learning that also serve to brand and differentiate schools in the education marketplace. The overarching economic imperative is evident in the high-profile provision of laptop computers throughout the state of Maine. The goal of the Maine Learning Technology Initiative (MLTI) is to: 'Transform Maine into the premier state for utilizing technology in kindergarten to grade 12 education in order to prepare students for a future economy that will rely heavily on technology and innovation' (cited in Silvernail & Harris, 2003, p. i).

Thus, many deschooling initiatives also support the goals of increased competition and deregulation, in keeping with the free market policies of neo-liberalism, which is the real paradox of decentralisation (Valovic, 2000). As New Zealand embarks on a similar policy agenda, it is timely to remember that it is simply absurd to think that capitalism wants a good many critical thinkers (Neill, 1995). The following quote reveals the extent to which the link between educational success and economic prosperity has taken root in the teaching profession. As Stephen Heppell (2001), director of UltraLab and a key advisor to New Zealand's Ministry of Education writes: 'Will our technologically induced learning futures show that we made a good use or bad of the opportunities presented? Future economic prosperity will tell us tomorrow' (p. xvi).

The key point is that the deschooling discourse harbours an inherent tension between endorsing neo-liberal market reforms and embracing an empowering paradigm that challenges the entrenched and oppressive nature of schooling. Although many deschooling initiatives have tremendous potential, they often embody a set of values quite different from education as a public good in which the government is responsible for the provision of a strong education system. Arguably, much of the deschooling discourse promotes a type of e-learning very distant from the original vision of a learning society committed to active citizenship, liberal democracy and equal opportunities for all (Faure, 1972).

The reconceptualist discourse advocates critical pedagogy and the promotion of education *for* democratic citizenship. This discourse encompasses not just knowledge of democracy but skills and understandings for active participation in all aspects of the knowledge society. Notably, the reconceptualist discourse is not driven by an ICT imperative or the libertarianism of the wired (Warnick, 2001), rather it is concerned that the ICT-related school reform movement is bypassing deeper questions about the nature of the knowledge society. The discourse recognises the ethical, moral and political dimensions of good pedagogy largely ignored in the new science of learning. Thus, the distinguishing feature of this discourse is the way in which e-learning is located in wider debate about the purpose and fundamental values of education. The discourse adopts the view that teachers need to educate students so they can not only participate as future citizens but also actively shape a very different kind of world. There is a basic need to reconceptualise the curriculum to promote an active citizenry capable of transforming society to ensure a more equitable education system and socially just future for all.

Finally, the socio-cognitive discourse advances new understandings of learning and teaching from contemporary developments arising out of the so-called cognitive revolution – that is, the new science of learning. The socio-cognitive discourse itself needs no further discussion, as it is already well-documented in the professional literature. The key point is that the other four discourses all draw on contemporary developments in cognition science, but with very different intentions. The language of the new ways of learning acts as a meta-discourse that simultaneously infuses the other discourses, which thereby renders the socio-cognitive discourse as problematic.

The metaphorical hourglass illustrates how the sands of the competing and coexisting discourses all funnel through educational concepts and theories from the new science of learning. Consequently, they are difficult to disaggregate from the socio-cognitive discourse. Put another way, contemporary developments in learning theory conceal many of the desired non-educational outcomes behind the ICT-related school reform movement. In this sense, the different interest groups and stakeholders borrow a socio-cognitive language of persuasion to legitimise their own hegemonic agenda. The result is that teachers can end up collaborating with the enemy (Hargreaves, 1995).

Conclusion

There are complex forces behind the drive to reform schools through ICT and, rather than be lured by the appeal of real-world learning, the teaching profession needs to create a culture of activism and reconceptualism. Such a culture would reclaim the true status of teaching in a liberal democratic society and raise teachers' political awareness and critical consciousness where pedagogical activism is a moral imperative. As Fullan (1993) writes: 'Moral purpose without change agentry is martyrdom; change agentry without moral purpose is change for the sake of change' (p. 14).

This chapter has shown that there is a key difference between education *in* change as opposed to education *for* change. In the context of the political nature of the digital landscape, teachers need to address the most basic questions of purpose and meaning. What kind of society do we want? What is the meaning of schooling in the knowledge society? What are the real problems confronting schools that need solutions? What conditions must policymakers provide for teachers if the public education system is to be fair and equitable? Such questions bring issues of critical citizenship, democratic community and social justice to the forefront of discussion.

References

Abbott, C. (2001). *ICT: Changing education.* London: Routledge/Falmer.

Abbott, C., Lachs, V., & Williams, L. (2001). Fool's gold or hidden treasure: Are computers stifling creativity? *Journal of Education Policy*, 16 (5), 479–87.

Apple, M. (2003). *The state and the politics of knowledge.* New York: RoutledgeFalmer.

Armstrong, A., & Casement, C. (1998). *The child and the machine: Why computers may put our children's education at risk.* Ontario: Key Porter Books.

Bowers, C.A. (2000). *Let them eat data: How computers affect education, cultural diversity, and the prospects of ecological sustainability.* Athens, GA: The University of Georgia Press.

Brown, M.E. (2003). Beyond the digital horizon: The untold story. *Computers in New Zealand Schools*, 15(1), 34–40.

Burbules, N. & Callister, T. (2000). *Watch IT: The risks and promises of information technologies in education.* Boulder: Westview.

Clegg, S., Hudson, A. & Steel, J. (2003). The Emperor's new clothes: Globalization and e-learning in higher education. *British Journal of Sociology of Education*, 24 (1), 39–53.

Codd, J. (2002). *Globalization and education: What does it mean for Aotearoa/New Zealand?* Paper presented at the Annual Conference of the New Zealand Association for Research in Education, 5–8 December, Palmerston North.

Codd, J., Brown, M., Clark, J., McPherson, J., O'Neill, H., O'Neill, J., Waitere-Ang, H., Zepeke, N. (2002). *Review of future-focused research on teaching and learning.* Wellington: Ministry of Education.

Cordes, C., & Miller, E. (2000). *Fool's gold: A critical look at computers in childhood.* Maryland: Alliance for Childhood. Available at: http://www.allianceforchildhood.net/projects/computers/ computers_reports_fools_gold_contents.htm

Cuban, L. (2001). *Oversold and underused: Computers in the classroom.* Cambridge, MA: Harvard University Press.

Faure, E. (1972). *Learning to be.* International Commission on the Development of Education. Paris: UNESCO.

Fullan, M. (1993). Why teachers must become change agents. *Educational Leadership*, 50(6), 12–17.

Gates, B. (1995). *The road ahead.* New York: Penguin Books.

Ham, V., & Wenmoth, D. (2002). *Educator's use of the Online Learning Centre (Te Kete Ipurangi) 1999–2001.* Wellington: Ministry of Education.

Hargreaves, A. (1995). Beyond collaboration: Critical teacher development in the post-modern age. In J. Smyth (ed.), *Critical discourses on teacher development.* New York: Cassell, 149–80.

Healy, J. (1998). *Failure to connect: How computers affect our children's minds – for better and worse.* New York: Simon and Schuster.

Henwood, F., Wyatt, S., Miller, N. & Senker, P. (2000). Critical perspectives on technologies, in/equalities and the information society. In S. Wyatt, F. Henwood, N. Miller & P. Senker (eds), *Technology and in/equality: Questioning the information society.* London: Routledge, 1–18.

Heppell, S. (2001). Preface. In A. Loveless & V. Ellis (2001). *ICT, pedagogy and the curriculum.* London: Routledge, xv–xix.

Hlynka, D. (2003). The cultural discourses of educational technology: A Canadian perspective. *Educational Technology*, 43(4), 41–5.

Kirkpatrick, H. & Cuban, L. (1998). Computers make kids smarter – right? *Technos Quarterly for Education and Technology*, 7(2), 1–11.

Lai, K.W. & Pratt, K. (2003). *Learning with technology II: Evaluation of the Otago primary schools technology project.* Dunedin: Community Trust of Otago, New Zealand.

Lai, K.W., Trewern, A. & Pratt, K. (2002). Computer coordinators as change agents: Some New Zealand observations. *Journal of Technology and Teacher Education*, 10(4), 539–51.

McMillan, B. (1993). Conference reflections. *Computers in New Zealand Schools*, 5(2), 3–4.

Ministry of Education (2004a). *Making a bigger difference for all students: Schooling strategy discussion document.* Wellington: Medium Term Strategy.

Ministry of Education (2004b). *Learning@schools: Shaping teaching and learning in the 21st century.* Available at: http://www.learningatschool.org.nz

Ministry of Education (2003). *Education statistics of New Zealand for 2002.* Wellington. Ministry of Education.

Ministry of Education (2002). *Digital horizons: Learning through ICT – A strategy for schools, 2002–2004.* Wellington: Learning Media.

Ministry of Education (2001). *Information and communication technologies (ICT) strategy for schools, 2002-2004 (draft).* Wellington: Learning Media.

Moss, S. (2002). *Towards a model of best practice for integrating ICT across the curriculum.* Occasional Papers, Research Information for NZEI Te Riu Roa Members, Wellington.

Neill, M. (1995). Computers, thinking, and schools in the 'new world economic order'. In J. Brook & I. Boal (eds), *Resisting the virtual life: The culture and politics of information*, San Francisco: City Lights Books. 181–94.

Oppenheimer, T. (2003). *The flickering mind: The false promise of technology in the classroom and how learning can be saved.* New York: Random House.

Oppenheimer, T. (1997). The computer delusion. *The Atlantic Monthly*, 280 (1), 45–62.

Peters, M. & Marshall, J. (2004). The politics of curriculum: Autonomous choosers and enterprise culture. In A-M. O'Neill, J. Clark & R. Openshaw (eds). *Reshaping culture, knowledge and learning? Policy and content in the New Zealand Curriculum Framework*, Palmerston North: Dunmore Press, 109–25.

Postman, N. (1996). *The end of education: Redefining the value of school.* New York: Random House.

Postman, N. (1993). *Technopoly: The surrender of culture to technology.* New York: Vintage Books.

Principals Today. (2004). *Principals Today*, Term 1, p. 28.

Reeves, T. (1998). 'Future schlock,' 'the computer delusions,' and 'the end of education': Responding to critics of educational technology. *Educational Technology*, 38 (5), 49–53.

Rosen, B. (1998). *Winners and losers of the information revolution: Psychological change and its discontents.* London: Praeger Publishers.

Roszak, T. (1994). *The cult of information: A neo-luddite treatise on high tech, artificial intelligence and the true art of thinking.* Los Angeles: University of California Press.

Selwyn, N. (2002). *Literature review in citizenship: Technology and learning. Report 3: Nesta Futurelab Series.* Available at: http://www.nestafuturelab.org/research/lit_reviews.htm

Selwyn, N. & Gorard, S. (2002). *The information age: Technology, learning and exclusion in Wales.* Cardiff: University of Cardiff Press.

Silvernail, D. & Harris, W. (2003). *The Maine learning technology initiative: Teacher, student, and school perspectives mid-year evaluation report.* Maine Education Policy Research Institute. Available at: http://www.usm.maine.edu/cepare/pedf/ts/mlti.pdf

Smyth, J., Dow, A., Hattam, R., *et al.* (2000). *Teachers' work in a globalizing economy.* London: Falmer Press.

Stoll, C. (1999). *High-tech heretic: Why computers don't belong in the classroom and other reflections by a computer contrarian.* New York: Double Day.

Telecommunications Users Association of New Zealand (TUANZ). (2004).

Valovic, T. (2000). *Digital mythologies: The hidden complexities of the Internet.* Piscataway, NJ: Rutgers University Press.

Warnick, B. (2001). *Critical literacy in a digital era: Technology, rhetoric, and the public interest.* Mahwah, NJ: Lawrence Erlbaum Associates.

Yelland, N. & Rubin, A. (2002). *Ghosts in the machine: Women's voices in research with technology.* New York: Peter Lang.

e-Learning and Pre-Service Teacher Education

3

Nola Campbell and Russell Yates

Introduction

According to the Ministry of Education (2002, p. 11) the 'amount of readily available information about how teachers include ICT within their classroom programmes is limited', as is information about how students use ICT in their learning. This chapter, based on the experiences of two lecturers in the Mixed Media Programme (MMP) of pre-service primary teacher education at the School of Education, University of Waikato, seeks to begin to address issues in relation to a group of pre-service teacher education students. It is an e-learning journey founded on the belief that flexible teaching practice that utilises ICT can provide new opportunities for teacher education students in remote areas of the University's region. The MMP is also a journey founded on a need to provide teacher education in areas where students and potential students were, and are, unable to attend regular classes on a university campus.

Previous distance education initiatives by the School of Education (formerly Hamilton Teachers' College) in the Gisborne/East Coast region in the early 1990s had been very successful but had involved staff driving from Hamilton to teach classes. Although successful in providing some teachers for rural areas, it was also tiring and time-consuming for staff. A cost-efficient alternative for the institution had to be found for a sustainable distance-teaching programme that could meet the needs of both staff and students. As Waikato University was the first tertiary institution in New Zealand to have access to the Internet, some teaching staff were already exploring the opportunities that the online medium could provide. Involvement in national ICT initiatives was highlighting the use of technology that could be embedded in a teacher's practice and based on real needs.

The development of MMP in 1997 was founded on existing experience in the online medium, coupled with continued concerns expressed by rural school principals in the central North Island region who, at that time, were finding it difficult to recruit and retain teachers. They believed that if teacher education could again be provided in students' own homes and communities, then there was considerable evidence from the previous programme that these graduates would be available to local schools once they had completed their degrees. This 'grass roots' request provided a unique collaborative opportunity between schools and the University and the chance to integrate e-learning as a vital part of this process.

The Mixed Media Programme

The Bachelor of Teaching through MMP was the first tertiary-level online degree available in New Zealand utilising the Internet and a mixed-mode approach to teacher education. It began in February 1997, with fifty-four students from across the central

North Island region arriving for a week on campus to begin their three-year Diploma (later Bachelor) of Teaching course.

The programme combines elements of flexible learning by utilising both online and face-to-face approaches. Each year, students attend three one-week long block courses on campus in February, June and August; they carry out tasks in a local base school and they communicate with lecturers and other students through the Internet in a learning management system now known as ClassForum. Just like on-campus students, MMP students undertake a period of teaching practicum in a school in their local area. The practicum involves the students being placed in a school as a teacher for a period of four or more weeks.

The compulsory papers required to complete the Bachelor of Teaching degree are the same as those for on-campus students, often with the same staff member teaching both online and face-to-face. A first-year MMP student will have a programme that follows the pattern shown in the following figure:

Figure 1: A typical year for an online MMP student

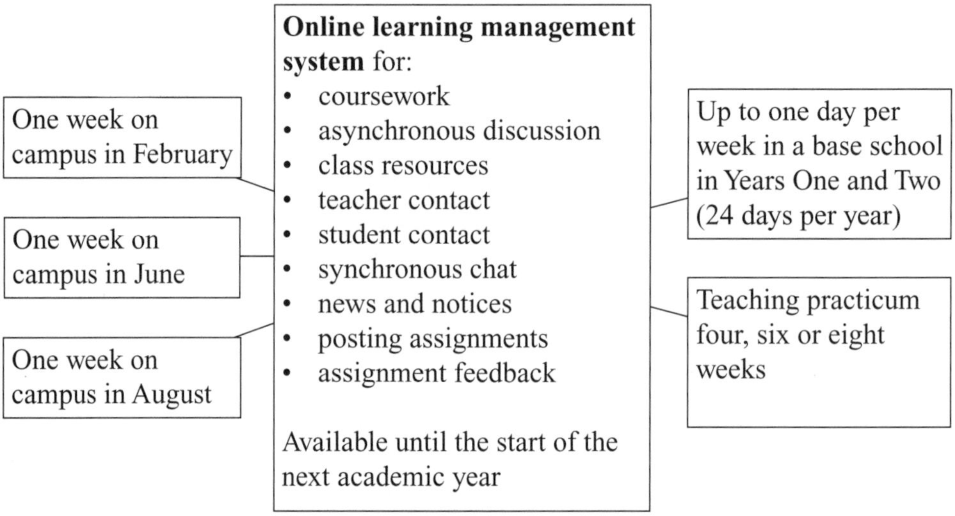

One aspect that is apparent in Figure 1 is the central nature of the online communication that takes place. Online communication for these teacher education students is through the University of Waikato platform known as ClassForum. It provides students with access to coursework and to a variety of both synchronous and asynchronous modes of interacting with lecturers and peers. Students generally have access to their online classes one week before they arrive on campus at the start of the semester in their first year. This enables them to come to their campus week having completed some preparation. This early access to online coursework ensures that the on-campus time is used effectively to clarify any uncertainties. Time on campus is spent meeting with lecturers, peers and support staff, so that a learning climate is established and maintained early in a student's academic life. Access to the learning management system, ClassForum, is ongoing until a student completes their studies.

Support systems for students

The Mixed Media Programme provides a variety of support systems. In addition to an online forum for each class he or she is enrolled in, there are a number of areas that the online student will have access to over the period of study. These include an area for students in each annual intake to talk with each other and to talk with the MMP administrator and coordinator. In this open forum, common concerns and questions can be aired with the whole group.

Another online facility in ClassForum is the Virtual Education Reference Desk (VERD). The VERD provides access to information and to the Education Library staff, many of whom have had experience as information coaches in some of the online classes so are very good 'cybrarians'. A number of cybrarians have completed their Master of Library Science online from Victoria University and have had invaluable first-hand experience as distant learners. They are therefore able to provide an effective conduit to the regular library.

Finally, ClassForum has a Guided Tour and an OnLine Campus offering support for staff and students about how to use the online environment. This facility relieves teaching staff of the task of explaining software issues and lets them get on with teaching their classes.

Selecting students for a successful online teacher education experience

Many tertiary-level students who enrol in e-learning opportunities are accepted into a course or programme of study based on meeting enrolment criteria, successful completion of application forms and payment of fees. There are generally few restrictions in terms of suitability in schools, colleges and faculties that are not preparing students to be members of a professional group.

Students who wish to enter teacher education, either online or on campus, are required to be selected following an application process. The MMP applicants undergo a shortlisting and interview process to ensure suitability for both the teaching profession and for learning online. The applicants for the programme are generally mature women who live in both urban and rural areas of the North Island. The majority have family commitments that make shifting closer to the university not an option for them. It was considered essential to select applicants who were not only suitable to become teachers but also to have some of the e-learning attributes that would enhance their completion of the programme. As e-learners are a new 'breed' of student, it is essential to explore the aspects that may impact on their success both as teachers and learners in virtual classroom environments.

When students have been shortlisted they are asked to attend an interview, which is conducted either face-to-face or by telephone. Support people are encouraged to attend interviews if the applicant is more comfortable in this situation. The selection interview is ideally conducted by two staff who are currently teaching online in the programme. This enables the interviewers to answer questions the applicant may have about the structure of the programme and the mode of interaction, as well as gain a picture of the applicant's suitability for e-learning. Staff members, who understand the challenges for online learners first-hand, are more likely to empathise with the demands of learning online and be able to point out some of the challenges. Staff are also able to share some of their experiences with online students, so that applicants are informed about what they might be 'letting themselves in for' over the next three years.

Some of the discussion during the interviews relates directly to the online mode of learning and relevant support systems and may include questions from staff like:

- Do you know anyone in your area who is an MMP student?
- What do you currently use your computer for?
- If you got into difficulties using your computer, where could you get help?
- If you got into difficulties with the online coursework, what could you do to get help?

As the applicant responds to these questions, the staff can ascertain how realistic they appear to be about teaching and learning online. The use of ICT as their vehicle to class can sometimes give the impression that the technology is the focus of what happens over the three years, when in fact it should be no more than the conduit for interaction. The use of ICT is just a means to an end.

Other aspects identified by staff may impact on the online behaviour of the student and can include a student's:

- ability to work in a collaborative way;
- ability to work independently;
- academic ability;
- ability to problem-solve and judge when to ask for help;
- motivation for applying to the programme.

The purpose of this careful selection is to attempt to avoid the high dropout rates and poor retention and completion patterns that have been seen in some other distance and online education programmes of study (Lockett, 2000). A small number of students who have not had the benefit of another MMP student within their region have struggled with their isolation, so this support factor needs to be considered and discussed with the applicant.

The most profitable and effective interviews have been those which are 'conversations' between the applicants and the interviewers, along with family members who have come along in support. These 'conversations' can enable the interviewers to gather a strong profile of the applicant and her or his support systems, while the applicant gains an understanding of the nature and demands of the programme. Students who enter the programme following a sound selection pathway are generally better prepared for successful study.

The Four Rs of e-learning

It is useful now to consider in more detail some of the key factors that will impact on the student's e-learning journey. Four attributes that will be important are relationships, reflection, resourcefulness and resilience.

Relationships

Communication with other students and staff and collaboration with peers are important aspects that are considered at selection for an online programme of study. These abilities are vital, both when students are on campus for such a short period and when they are working online.

When students are on campus for the first time, a number of activities are specifically aimed at assisting them to establish relationships, which will assist their online learning experience. Many of the students choose to stay in the Te Kohinga Marama Marae when on campus, and this accommodation choice provides them with a very strong 24-hour learning network of peers. It has been found that by undertaking relationship-building arrangements and activities, there is a better chance of student retention, as they will be more likely to seek one another's support, particularly in times of difficulty.

Reflection

Learners who choose to learn in non face-to-face situations are required to work in a reflective manner, an unfamiliar experience for some people. Many applicants do not have a history of successful academic achievement so their ability to become reflective learners is often difficult to gauge. The opportunity to be an online teacher education student in the MMP has often been seen as a 'second-chance' following limited success at school. Involvement in community groups, committees and other courses of study may give a further indication of reflective learning potential.

Some students have, prior to their teacher education application, already completed one or two courses as a way of proving to themselves and the University that they are capable and reflective students. Often this activity is a clear indicator of both academic ability and a commitment to succeed.

Resourcefulness

Although the MMP online programme utilises a range of media approaches, the key element to success is an ability to operate and communicate through a computer. The resourceful learner will quickly learn that you do not have to know everything, you just have to know who does. Knowing when to ask for help and how to ask for help are sometimes new experiences for e-learners.

In 1997, many students began their online teacher-education degree with little or no prior computer experience, some acknowledging they did not even know how to turn the computer on but were keen to learn. In these initial stages it was considered that teaching students how to use a computer would be an integral part of the programme. However, student social networking quickly enabled less confident students to succeed. Within approximately eight weeks, very few of the students found using a computer difficult. They commented that they needed to use the computer as a tool and therefore they needed to learn to use the tool quickly.

To gain a greater understanding of the technological challenge faced by some MMP students, Donaghy *et al.* (2004) undertook a small study entitled 'The Impact of Information and Communication Technology upon Schools'. Eleven students were interviewed about their level of ability and experience before entry. Collectively, the group had little or no experience prior to entry but soon gained both confidence and skills: 'It was like a whole new world opening up, in terms of confidence building. I mastered the basics of the technology side of it and it is now part of my school programme' (p. 5). Clearly, when the focus of the e-learning experience is on the teaching and learning, then the tools become less feared and less visible.

Additional support resources in the form of brochures, CDs and videos have all been used to assist students, but nothing has been as successful as the peer mentoring and support. This is fortunate because time spent on campus should not be wasted learning how to use a computer. Regular on-campus students who travel to the institution by car do not receive driving lessons to enable them to reach the campus car park. Similarly, online students who use the Internet to travel to their online class should not expect lecturers to give them basic computer 'driving lessons'.

Many MMP graduates reported that they were seen as a valuable resource in schools because of the way in which ICT had become part of their own everyday learning experience, a personal journey that would become a professional asset for the future.

Resilience: Motivation to succeed

A strong desire to succeed in an online programme of study is necessary to sustain the long and lonely hours of reading, reflection writing and online discussion. Students need this strong motivation to succeed and to have a range of problem-solving strategies for when they need help. Time-management skills and an ability to work unsupervised are driven by a strong motivation toward success and are key attributes that are critical for online learners.

Unexpected events such as illness, flood, pregnancy, part-time work, housework and daily living all challenge the resilience of the e-learner.

Lessons from the e-Learning Student Selection Process

The students who have been isolated in rural regions with nobody to call in and share a cup of tea and a chat are likely to experience the effects of distance-learning isolation more than those students who are selected in regional clusters or groups. Lockett (2000) found that isolation was a critical factor in the motivation and drop-out rates of students. She described the value of an online campus as a means of students feeling a sense of belonging to an institution.

It has been our intention to encourage independence and self-reliance among students so that teaching staff can focus on the teaching activity and not be the sole supporters for students who are feeling separation anxiety. For this reason, when selecting applicants for interviews, it is important to identify the area in which they live and look for other applicants who are 'just down the road'.

Our view that the Internet is a student's vehicle to class is supported by the fact that it is stated clearly that this is a requirement of entry to the MMP. Some online students with previous traditional distance-learning experiences have described how, when they received piles of material via regular mail, they followed a set pattern with little or no interaction with the teacher and then mailed back their work: 'they told us what to do and when to do it, we did not really have to think for ourselves as everything was provided.' In contrast, the online learning experience highlights teaching and interaction, so students are no longer in a situation of teaching themselves but are expected to join a community of online learners. They have to take a measure of responsibility for their own learning, which is often a very new experience. Some students succeed with excellent results and others, who find time-management difficult, will fall by the wayside.

Revolution or Evolution?

It could be argued that e-learning and ICT have produced a revolution in pre-service teacher education. The reality is a little more like an evolution. Institutions were no longer able to ignore the very tools that were becoming commonplace in the community and workplaces. People would be using these tools anyway to complete a variety of activities in their daily lives, so to ignore them as tools for learning would be to bury our heads in the sand. Certainly, there have been people who have adopted this ostrich-like behaviour but frequently they have not had the opportunity to see how the technology could enhance what they already are doing – for example, the ability to email distant friends and family has given many adults and teachers the chance to explore the potential of an online environment. This level of ownership of change is essential if any evolution is to amount to a revolution in the way we teach and learn.

In order to understand this r/evolution, it is useful to look at the impact of online learning on the students, staff, institutions and, most importantly, on the schools where the pre-service and graduate teachers are working.

The Impact of e-Learning on the Student's Learning Experience

From the outset of the MMP programme, students have been clear in acknowledging the part that ICT plays in their study and in their lives generally. Many students who have successfully completed the programme began as ICT novices. Their ability to use ICT has now transferred into their use of ICT in classrooms, to the extent that it is part of their everyday use.

Students are very clear that the online experience is important for them. Anecdotally they recognise that it is the key to their ability to gain a qualification. As they have become more familiar with online experiences, so too have they become more aware of the need for quality. That is an implicit thread in Donaghy *et al.* (2003) who reported on student experiences in MMP. In this study, students commented on the way they organised their time for online study, the support systems they used to maintain their online study, and the nature of the online environment, including feedback, discussions, interaction with their tutors and the layout of course materials.

For most students, the online experience is satisfying. One of the keys to successful online learning is the way that they organise their time. Since MMP began operating, it has become clear that this is one of the factors that influences successful learning and, implicitly, student retention. Donaghy *et al.* (2003) investigated this aspect and reported that: 'they use wall planners, calendars, spreadsheets and diaries to manage their time. Their actual study time was divided into course readings, checking ClassForum regularly, reading and contributing to discussion and doing assignments. Family commitments and paid work were influential factors as to when students studied, but it was assignments, modules and discussions which determined how they used their time.' (p. 9). The range of approaches is significant and emphasises the need for students to be resourceful and reflective. They need to think through their own circumstances before embarking on online learning because the time requirement is critical.

Students in online programmes readily develop support systems that help them ensure survival in the online environment, and are encouraged to do so. In this particular programme, their sources of support are varied. They include other MMP students, their families, their local school and the teachers they work with, the University administration, friends and a range of online techniques including live messaging (Donaghy *et al.*, 2003, p. 13).

The final aspect to consider when discussing student responses to online learning is their online environment. The MMP students have always used ClassForum as their environment and have been clear in what they expect from it. They expect that materials are available at all times, that course materials are clearly laid out, that instructions are clear and unambiguous. They also report regular, specific feedback is critical to their learning. Indeed, many responses from students indicated that this may be the most important aspect. This is understandable when it is considered that many online students are those who are returning to study following either a long absence from it or earlier unsuccessful school experiences, or both of these factors. This positive comment on feedback exemplifies the points students made to Donaghy *et al.* (2003) about good feedback when they said 'Lecturers provided excellent feedback – prompt, specific, constructive and positive' (p. 25).

The Impact of e-Learning Graduates on Schools in the Region

When Murray and Campbell (2000) examined barriers to implementing ICT in 532 New Zealand schools, they identified the five main obstacles as: teachers' limited skills; funding; lack of teacher training; time/workload/curriculum pressure; and lack of equipment (p. 4). When the MMP programme was mooted and then developed, it was thought that there would be a significant impact upon the ICT life of schools involved in the programme. However, the impact of online pre-service students and graduates in schools in the region has been varied. On one hand, it is clear that the level of expertise of individual students has increased significantly during their teacher education and that this has impacted on individual teacher programmes. On the other hand, it is not so clear that there has been such an impact on school-wide programmes. There may be various reasons for this but the involvement of schools in national ICT programmes and the general school-wide focus on ICT appears to have also been significant. What e-learning graduates have done is to begin their teaching careers with many of the ICT skills, attitudes and resources that they will need to assist the school to move forward. Donaghy *et al.* (2004) reported how one school's deputy principal had acknowledged the role of MMP students in the school and said: 'They are the first ones to raise their hands for further training because they realise the importance of, not only ICT, but also ongoing ICTPD' (p. 28).

Donaghy *et al.* (2004) also commented on several school leaders who felt that graduates who had been e-learners were better prepared in terms of using ICT in their teaching and that this factor was important during the school's selection and employment process.

The Impact of Online Teaching on Academics

Academic colleagues have not always been positive toward teaching online and the MMP. Initially, the need to find staff to teach classes for all papers in the degree did mean that the opportunity to teach online was not always seen in a positive light. The move meant that many staff were being asked to step outside their comfort zone and take some real risks for what, in 1997, were uncertain gains.

Asking academic staff to teach in a way they never learned has prompted some interesting responses. These have ranged from extreme anxiety about how to teach effectively in a relatively foreign environment to a belief that, somehow, once the coursework is online then teaching and learning will somehow magically happen without their intervention. Clearly, staff support and development is essential and tertiary institutions across New Zealand are trying a variety of models. What defines the type and level of support is usually the balance (or imbalance) between technical skill development and supporting an understanding of an e-learning teaching pedagogy.

One effective staff-support measure that was implemented in some online classes was the concept of team teaching. This enabled new staff to be mentored into the new mode of teaching, while having the security of a colleague to guide them. It also provided a measure of accountability for the institution and the assurance for students that they would not be left alone without any staff contact.

The Impact of Online Teaching on the Teacher Education Institution

The integration of online teaching into a teacher education programme has provided the impetus for a thorough unintentional review of many on-campus classes. As each class has been considered for teaching online, a careful examination of the objectives, resources and activity often revealed that what had been done in the past on campus was sometimes based on custom and practice rather than sound educational reasons. Such questions as; 'Do you really need that much assessment?', and 'How does this topic fit with the goals of the course?' often prompted confused responses from staff who clearly valued this in-depth review of what they had been doing as it enabled them to reflect on new and different options.

Once the MMP students were reaching their final year of study they started to ask 'where to from here?' We had promoted the notion of lifelong learning but were not offering these pathways. The students asked us to make available graduate-level study opportunities online so that they could continue into Honours and Masters programmes of study.

With the successful online primary teacher education programme underway, it was not long before secondary teacher education was considered as an online alternative. Several small groups at the School of Education have been involved in this development, but to offer a fully MMP option has proved difficult because of the need to provide for a wide range of subject areas.

With the establishment of MMP in 1997, it was not long before other schools and faculties in the University were showing interest in what the teachers in the School of Education were doing. After seven years of the MMP, it is interesting to note that in 2004 every paper in the University has an online presence.

The changing role of the library and the library staff in response to e-learning has been a significant development in collaborating with academic staff. Initially, students who did not attend regular on-campus classes posed quite a challenge for the library because existing custom and practice simply did not work for this new group of 'clients'. Perrone (2000) described this change in a positive manner: 'There are significant differences between the traditional on-campus and online teaching environments that enable librarians to expand their roles to provide more effective information literacy education' (p. 4).

This opportunity for librarians to be part of the teaching team was warmly accepted by a growing and enthusiastic group of librarians and this was evident in the development of information coaching (Campbell, 1999). Perrone described how: 'The term "information coach" was adopted as it emphasized that the librarian's task is to "coach" students in finding and using information for themselves, whereas for many people (particularly those less familiar with academic libraries) the term "librarian" is associated with someone who finds information for others.' (p. 3). Information coaching provides some equity of service for online students by combining library teaching with a reference service. It is a new way of working and the inclusion of a librarian as part of the teaching team has advantages for all parties involved – librarians, students and academic staff. For students, the primary benefit is an increase in their level of information literacy and the opportunity to see librarians as learning partners.

Conclusion and Discussion

When forty-eight students graduated in April 2001 from the 1997 cohort, there was a great sense of pride. Families came to celebrate and share the occasion and schools celebrated their own role in assisting the education of these new teachers. This would not have been possible without a combination of factors that involved both people and technology.

The online pre-service teacher education programme has changed teaching and learning for many staff and students. In many ways it has been seen as a change for the better and one that is clearly now irreversible. There is no going back to a pre-ICT era. Besides, who would want to when they have had their own successful experience?

As time passed, the MMP was to have a significant impact on the way online teacher education was to evolve both locally and nationally. Regular on-campus students at Waikato were wanting to 'attend' some online classes and graduate students were wanting online Masters' degrees so they could continue their academic careers while teaching during the day. Other teacher education institutions, such as Christchurch College of Education with its Primary Open Learning Option (POLO), began to utilise online technologies, and Wellington College of Education, with its graduate distance Learn Online programme, began to offer a variety of flexible approaches to teacher education.

Future Challenges

The scalability of online teacher education and its ability to maintain a quality experience have to be major challenges for the profession. How do we maintain a learner-centred experience for the student that models good teaching practice for future teachers? How can we expect staff to know how to teach online when they have not learned in this way?

Another interesting issue is how we can move from the successful boutique ventures seen in many institutions to a larger programme of online teacher education that is sustainable. The opportunities for collaboration across institutions is encouraged at a national level but rarely seen in practice. An emphasis on collaboration across online teacher education programmes could see a strong national profile for this mode of teaching and learning.

As society continues to accept the Internet as part of daily life, then it is reasonable to assume that e-learning experiences at school will transfer into expectations of tertiary education. It is no longer a question of whether or not we should continue to educate pre-service teacher education students online but how we will actually go about doing this and maintain the values and standards expected of members of the teaching profession. In terms of the future, Jones (2002) argued, 'it might be expected that all entrants into pre-service teacher education courses will have mastered these basic computing skills' (p. 6). Kendall (2003) emphasised how 'learners now have a choice and increasingly wish to combine options, choosing when and where they study and learn' (p. 1).

One thing is certain, the technology and what we can do with it will change on a regular basis. As teacher educators we need to ensure we are not lured by the new technologies down pathways that prove both fruitless and expensive. Barriers imposed by the technology evaporate when we focus on our goals of quality teaching and learning that model the way we expect our teacher education students will work in their own classrooms and schools.

References

Campbell, N. (1999), Team teaching in a web based classroom. In *Open, flexible and distance learning: Challenges of the new millennium: Collected papers from the 14th Biennial Forum of the Open and Distance Learning Association of Australia.* Geelong: Deakin University (45–49).

Donaghy, A., McGee, C. & Yates, R. (2004). *The impact of information and communication technology upon schools.* Hamilton: Wilf Malcolm Institute of Educational Research.

Donaghy, A., McGee, C., Ussher, B. & Yates, R. (2003). *Online teaching and learning: A study of teacher education students' experiences.* Hamilton: University of Waikato.

Jones, A.J. (2002). *Integration of ICT in an Initial Teacher Training Courses: Participants' Views.* Paper presented at the Seventh World Conference on Computers in Education, Copenhagen, July–August, 2001.

Kendall, M. (2003). *ICT and the Teacher of the Future.* Paper presented at the IFIP Working Groups 3.1 and 3.3 Working Conference: ICT and the Teacher of the Future, held at St. Hilda's College, The University of Melbourne, Australia, 27–31 January 2003.

Lockett, K. (2000). *Online support: Addressing the loneliness of the long-distance learner.* Paper presented at the DEANZ Conference Dunedin, 27–29 April, 2000 (218–25).

Ministry of Education (2002). *Information and communication technologies (ICT) Strategy for schools. Draft, 2002–2004.* Wellington: Ministry of Education.

Murray, D. & Campbell, N. (2000). Barriers to implementing ICT in some New Zealand schools. *Computers in New Zealand Schools*, 12(1), 3–6.

Perrone, V.G. (2000). The changing role of librarians and the online learning environment. In *Proceedings. Distance Education: An Open Question?* Papers presented at an international conference sponsored by the University of South Australia with the International Council for Open and Distance Education (ICDE) University of South Australia, Adelaide, 11–13 September 2000 [CDROM].

Clustering Schools for Teacher Professional Development in ICT: The 'Grand Experiment'

4

Vince Ham

What Makes for Effective Teacher Professional Development?

Globally, over the last decade, there has been a publicly perceived need, and a seemingly insatiable teacher demand, for professional development (PD) in all aspects of ICT use in education (APEC, 2004; Baldwin & Sinclair, 1994; Ministry of Education, 2001; Ofsted, 1994; Pelgrum & Anderson, 1999; Watson, 1993). The literature on the impact of these new technologies on schools abounds with descriptions of innovative projects claiming to exemplify 'good practice' in integrating information technology (IT) into teachers' classroom programmes, and many of these studies report some form of professional development activities or programmes as a catalytic intervention in the facilitation of such good practice. However, despite more than a decade of experience with PD programmes in many countries, at both national and local levels, the issue of what constitutes 'effective' practice in those programmes – what, specifically, works and what doesn't in terms of promoting quality use of ICTs for teaching and learning – is still a matter of considerable debate.

In a synthesis of research on teacher professional development in Britain, Bolam (1997) argues that there is general consensus that good professional training courses are characterised by: collaborative planning; a clear focus on participants' needs; careful preparation; an orientation towards practice and action using methods such as action learning, action research, experiential learning and quality feedback; a 'sandwich' structure alternating in-class and out-of-class activities; a careful debriefing; and provision for follow-up support in classrooms.

Joyce and Showers (1988), writing from a US perspective, tend to adopt a less structural, more procedural or interactional view of effective in-service. They distinguish primarily between 'training', which is likely to take place outside the normal routines of a school, and 'coaching', which is much more collegial and integrated with a teacher's normal classroom activities. They argue that the learning of new teaching can best be achieved if five elements are present in the training programme: the presentation of underlying theory and description of any new skill; modelling or demonstration of the skill; practice in the skill in a simulated course setting; feedback in this setting; and, most important of all, one-to-one coaching when new skills or knowledge are applied on the job.

In a specifically ICT context, McDougall and Squires (1997) argue that teacher development programmes in IT could most usefully be evaluated upon two criteria. First, 'good' PD is 'comprehensive', meaning that it should cover more than just personal skills in IT, and should include pedagogical issues such as the integration of IT into existing curricula, changes in teacher roles, and underpinning theories of education. Second, 'good' PD is 'authentic', meaning that it is seen to be relevant to teachers' day-to-day problems. In another Australian longitudinal case study, Johnson

(1995) concluded that opportunities for reflection, collaboration and personal challenge are important process variables. In line with Joyce and Showers, the modelling of successful teaching strategies, relevance to curriculum, and a focus on metacognition and student process not product are also advocated as important content for effective professional development activities.

At the level of broad principles, therefore, there seems to be some developing consensus. But the literature shows there is far from consensus about which specific operational models of in-service are the best at exemplifying these principles.

The PD programmes in IT reported in the literature employ a veritable dolly mixture of tactics. These include: short courses and workshops held in or out of the school, focused on either technical facility or the 'modelling' of good classroom practice, or both; the temporary appointment of in-school facilitators to work alongside teachers in classrooms; extended courses leading to formal qualifications held in local universities or teacher centres; the establishment of electronic networks to link teachers to each other and/or expert facilitators to discuss common problems (e.g. Brown, 1993; Davis, 1997; Kerr, 1990; LeBlanc, 1996; Norton & Sprague, 1996; Ofsted, 1994; Persichitte & Bauer, 1996; Rhodes & Cox 1990; Stanley, 1995;), and action research projects by individual teachers, or small groups of teachers, often working in collaboration with academics or facilitators from a local research institution (Davis, 1997; Johnson, 1995; Somekh, 1997). Moreover, for every study saying that a programme of short withdrawal courses seemed to work (Faseyitan *et al.* 1996; Knight & Albaugh 1997), there seems to be a contradictory one saying that electronic networking, or graduate programmes in universities, or employing a school-based facilitator worked just as well (Jones, 1997; Kellogg, 1996; Norton & Sprague, 1996). And for every one of these apparent success stories, it seems, there is another bemoaning the apparently geological pace of technological uptake and maintenance in education compared with the business community (Cuban, 1993, 2003; Kerr, 1990; Wiburg, 1994).

The particular virtues or vices of each model of in-service provision as mechanisms for providing such comprehension and authenticity are therefore still in dispute and the result is a confused and contradictory picture about what it actually is that is, or is not, making a difference for teachers. Despite Rhodes and Cox's (1990), and Ridgway and Passeys' (1991) attempts to do so in the early 1990s, it is still not possible to say with any conviction that any one particular organisational model of delivery is likely to be more successful than any other. All these various in-service 'models' have their advocates, and nearly all claim to have had an effect.

But if this is so, then the clear implication is that the outer, organisational 'form' of PD is intensely problematic as the best or only descriptor, or predictor, of later classroom 'effect'. It is arguable, for example, that in ICT the central problem lies in the content and the presumptions such courses make about needs, rather than in structural issues regarding their length, their location or whether an 'outsider' or an 'insider' is taking them (Rhodes & Cox, 1990). As Fullan (1996) and others have argued for some time, the keys to teacher change lie in deeper, perhaps more dynamic and more personalised, variables related to the discursive, institutional contexts in which the programmes occur, and in their interactional and interpersonal progress as socio-professional events and processes. In this regard, therefore, there is a continuing need to distinguish more clearly between 'modes' of in-service provision, meaning the methods of implementation and delivery used, and 'models' of in-service provision, meaning the overall 'purposes and focus of the INSET: how teachers are perceived and the ingredients of the development process' (Owen 1992, p. 128).

This chapter outlines the main findings of a project evaluation that attempted to combine the much written-about structural/organisational aspects with the less written-about interactional/dynamic elements of programme evaluation in relation to PD in ICT. It is found that while 'ingredients of the development process' are indeed distinguishable in ICT, they do not form into a convenient 'recipe' of 'best practice'. Such ingredients draw from the broadest possible range of key domains, including the social and affective, and not just the structural or organisational, or even interactional.

The Genesis of the ICTPD Cluster Programmes

Since 1991, successive governments and many schools in New Zealand have invested large amounts of money on new computer-based ICTs. Inspiring this investment have been a widely held public perception that developing technological competence has become an essential component of schooling in the information age, and an equally widely held professional perception that such new technologies have an important role to play as tools for teaching and learning across most aspects of the school curriculum (APEC, 2004). Proponents of both imperatives, and not least teachers themselves, have also tended to see teacher PD as a vital factor in achieving such goals. and as a result it has been the policy of successive New Zealand governments since 1991 to commit public funds extensively to the provision of PD programmes for teachers in this area (Sallis, 1991).

For much of the decade after 1990, teacher PD was delivered via centrally funded private and public providers offering Ministry of Education sanctioned modes of PD programmes to schools and teachers in their regions. The programmes were usually based on an 'expert consultant' or 'advisory' model of delivery, and were contracted out to full-time facilitators or consultants in IT, initially based in colleges of education or universities, but increasingly over the period contracts were also given to private providers. The basic forms of the programmes were centrally dictated, usually involving support for a teacher or group of teachers over time periods of up to six months, and increasingly with a requirement to concentrate on whole-school or whole-department development. These 'Ministry-contract' programmes as they became known, were granted on a regional basis for year-long programmes, aimed at a systematic coverage of all regions and school levels over time.

The contestable 'Ministry-contract' programmes were not the only form of PD available to schools or teachers, however. Nor were they the only source of government funding for teacher PD. The Ministry continued, for example, to fund advisory services around the country to employ IT advisors, whose services were made available to schools at no cost. There was also a notable growth of teacher enrolments in formal papers and courses in educational computing and IT, leading to diplomas and degrees from universities and colleges of education. Similarly, several private-sector initiatives were established in which individual schools were resourced to employ full-time resource teachers to coordinate the IT facilities of the school and to deliver an extensive school-based teacher development programme.

Such a 'shotgun' approach to PD, involving a variety of modes of provision, continued into the mid-1990s when the Ministry, urged strongly by influential business-based lobby groups, also began looking at both greater direct school 'ownership' of their teacher development programmes and the delivery of much more in-service 'over the Internet'. These were seen as potentially more effective ways of funding PD in IT than the regional-advisory programme model which had dominated till that time (ITAG, 1998). In 1996, for example, a system of contestable professional development funding was established, known as ITPD funding. Schools were encouraged to bid for such funding in groups or as individual schools, although most of them in fact bid as individual schools. The Ministry also began sponsoring the development of extensive 'professional development' websites for various curriculum subjects, beginning with English Online in 1997.

In 1998, however, most of these various, and perhaps somewhat *ad hoc*, initiatives were given some coherence through the announcement of a new national ICT strategy for schools (Ministry of Education, 1998). For the first time, there existed a national policy statement outlining both a general vision or mission for ICT in New Zealand education, and a series of mid to long-term infrastructural and PD initiatives that the government would support over time. One key initiative announced in the July 1998 budget was a programme of devoluted and contestable funding for PD, which became known as the twenty-three ICTPD School Clusters project.

This programme represented a significant departure from the policies and practices of the previous decade, and it has dominated central funding for PD in ICT since. The clusters project departed from previous PD modes in four key respects:

- First, the funding for teacher PD in ICT was to be devolved directly to schools which would act as both 'producers and consumers' of their own PD programmes, rather than being open to tenders from traditional 'providers' such as the colleges of education advisory services who then invited schools or teachers to participate.
- Second, the programmes were available only to *groups* of schools, which had committed to a 'clustered' model of professional development for the benefit of teachers in all the participating schools.
- Third, the programmes were to last and be funded over three years, much longer than the one year that had been the custom in the past.
- Fourth, no particular delivery model was mandated in the contracts themselves. Applicants for ICTPD cluster funds were expected to develop and propose their own models of delivery, rather than to implement a variation on a predetermined, Ministry-approved model as had tended to be the case in the past.

Late in 1998, applications were invited from schools around the country to cluster together to provide these PD programmes for teachers. The programmes were to focus on the integration of ICTs into a variety of teachers' professional practices, including increased use of ICTs for teaching and learning, increased use for school administration and the development of school policies and plans for ICT.

The basic structure of the ICTPD cluster programmes was centrally prescribed. A lead school – often, but not necessarily, one with a reputation for best practice in the area of ICT use – would form a collaborative partnership with other schools for the provision of up to three years of teacher PD in those schools. Each cluster was to receive $100,000 per annum in central funding. The funds had to be spent on teacher PD, and could not be used to defray schools' hardware, software or infrastructure costs. Beyond that common brief, however, schools were free to group themselves as they wished, and were encouraged to develop and propose their own models and modes of delivering their programmes.

Early in 1999, twenty-three such ICTPD school clusters in various parts of the country were selected under the scheme and began implementing their programmes. Since then the cluster programme has become the major focus of the government's provision of teacher PD in ICT and a cornerstone of successive national strategy documents (Ministry of Education, 1998, 2001, 2003). A further twenty-eight clusters began their three year programme in 2001, followed by twenty-two more in 2002, twenty-one in 2003, and, most recently, forty in 2004. As at March 2004, approximately half of the schools in the country are, or have been, involved in an ICTPD cluster programme.

The Research Project
The rest of this chapter outlines the key findings to date of an ongoing research project commissioned by the Ministry of Education to evaluate the effectiveness of the ICTPD clusters projects on a national basis. It draws extensively on results from evaluation reports on the first two cohorts of clusters, which ran from 1999 to 2001, and 2001 to 2003 respectively. The chapter is based on the author's two reports from which excerpts are reproduced here with permission. The first of these reports has been published by the Ministry of Education (2002) and the second has been submitted to the Ministry.

The general strategy used for the evaluation was a mixed method investigation of the operational nature and in-school effectiveness of the PD programmes offered in both the 'twenty-three' and 'twenty-eight' school clusters over the years 1999 to 2003.

Certain data were gathered from all of the fifty-one clusters that took part in the two programmes. These included: pre- and post-surveys of participants; interviews with key stakeholders, participants and Ministry officials; and various official documents such as cluster proposals and milestone reports. Additional data on PD events and on teachers' and students' subsequent use of ICTs in schools and classrooms came from over 200 participant interviews and over 400 classroom observations in various cluster schools. These interviews and observations were conducted in ten case-study clusters from the 1999–2001 ICTPD Cluster programme, and one case-study cluster from the 2001–2003 programme (Ham *et al.*, 2002, Ham *et al.*, 2004).

A list of ingredients, but no recipe for best practice

What then, makes for an effective ICTPD cluster? In the words of the report on the first (1999) cohort, there is no 'fixed recipe', no 'one size fits all' formula, and no 'best practice' model that aspiring clusters should replicate or current clusters adapt to (Ham *et al.*, 2002). As has been found in similar studies of many other PD programmes in contexts other than ICT, the relative effectiveness or ineffectiveness of a particular programme was an intensely situated and contextualised phenomenon, unique in many ways to each cluster. Certain features worked very well in some clusters and less well in others; certain objectives were more strongly pressed in some clusters than in others, and so on.

But the lack of a replicable 'recipe' does not mean that the studies have not provided some fairly clear indicators and, what is more, relative consensus among participants and stakeholders, as to what should be on the list of ingredients. How clusters selected, measured out and combined these ingredients, and how influential each of them was subsequently felt to have been in the overall taste of the 'meal', varied according to clusters', schools' or teachers' specific needs and situations. But there was relative consensus as to what those core ingredients *were*. Some of these ingredients are organisational, to do with cluster size and distribution, social demography, location, and so on. Some are issues to do with the content of the programmes and the nature of the activities involved. However, even more important seem to be a variety of human factors, such as the abilities of the facilitators and managers, how well the programme deals with the affective needs of participants, how much it allows for professional sharing and partnership, and so on – how well, in short, the cluster operates as a social/professional unit.

Size matters

The number of schools and the number of teachers involved in the cluster programmes varied considerably. The smallest clusters provided ICTPD programmes to between four and seven schools, and some fifty to seventy teachers. The largest clusters sought to serve up to forty-five schools and a potential teacher population of several hundred. The milestone reports provided by the clusters at these extremes, as well as the reports of the national coordinator of the programme, however, make it clear that while 'small' did not automatically mean 'more effective,' 'too large' did tend to mean 'less effective'. The guidelines given to the group of applicants in 2001 suggested that the experience of the first cohort shows that, within the existing budget and time constraints of $300,000 per cluster over three years, an optimum cluster size could be anywhere between five and twenty-five schools, but, more importantly, would involve a maximum of not much more than a hundred teachers.

The main perceived advantages of the smaller clusters were the greater opportunity to build a sense of cluster community, a greater ability to involve the 'whole school' and thus for the programme to affect school culture and not just individual classroom cultures, and increased time for facilitators to provide regular and ongoing support to individual teachers and establish sustainable relationships with them.

There was considerable consensus among stakeholders with regard to the importance of clustering by geographical locality. In some measure, this relates to a concept of 'natural clustering', in that schools which were close to each other were often seen as having a natural commonality of interest. This common interest came by virtue of serving the same or bordering communities in the large cities, or the entire local community in the case of smaller centres and towns, or the local cultural community in the case of the Kura.

Largely, however, the advantages of localised clustering were based in the pragmatics of providing PD services. If not inherently making the PD programme 'better', there was a clear feeling among participants in both cohorts that it at least made it more manageable. Facilitators, in particular, felt that they were more accessible to geographically grouped schools, spent much less time and money travelling, could more easily arrange combined-school workshops and practicums, and spent more time in individual schools. These were valued not just in themselves but because they were felt to allow facilitators to build an ongoing personal and professional relationship with school leaders and especially individual teachers.

In several of the first cohort of clusters, the selection of schools became more and more consciously focused on the local area close to the lead school over the period of the programme. Even in those clusters which of necessity dealt with schools spread over a wide area, the fact of that spread was consistently reported in evaluation interviews and surveys as a limitation on the most effective operation of the cluster. This was overcome only by dint of extra workload on the part of the facilitators and/or by reducing the number of schools or teachers that could effectively be supported in the time given.

Frequency matters more than size

Further 'ingredients' in creating a successful ICTPD cluster related to issues of time and timing. In particular, it was felt to be important for facilitators to have regular contact with participating teachers over the longest possible period of time. For facilitators especially, the importance of *regular* contact was important; for teachers, the main issue was increasing the *amount* of time they had to devote to ICT issues.

The most important timing factors, especially from the teachers' perspective, were the amount and frequency of release time available to them to attend PD events. Clusters varied in the amount and frequency of release time given to individual teachers for PD. In some clusters, an entitlement was built into the cluster model; in others, it was expected that teacher release was a contribution made by the schools; in others, no release time was given but the schools or the ICTPD contract arranged an extensive programme of before-school or after-school workshops, 'techie breakies', and so on.

Teachers in particular were very clear in their evaluations that the *amount* of time made available to attend such events and to maintain impetus and development were key factors for them. In the case-study clusters, teachers willingly took up the opportunities offered in pre- and post-school hour sessions, but it is clear from the evaluations that they valued even more any extended periods of release from classroom duties to attend workshops, seminars, conferences, and the like. In teachers' written evaluations, lack of time was seen as the most significant constraint on their further development with ICTs. For both cohorts, over half of the teachers felt this to be 'a significant concern' at the end of the project. Similarly, over 80 per cent of teachers in both cohorts regarded release time as one of the most useful PD strategies employed by the clusters.

Duration matters more than frequency

It is important to note here that a key timing issue for the clusters was not so much the length of time a particular *school* was involved in the programme, but the length of time an individual *teacher* was involved. This period differed considerably from cluster to cluster. Some teachers, for example, were involved over all three years of funding; others took part only for a few weeks or months. Some other clusters had individual schools or individual teachers take part in the programme for up to one year only, and then replaced those schools or teachers with a different set of participants each year thereafter. In effect, therefore, the latter model was to run one-year programmes three times for different sets of participants, rather than a 'three-year model' with the same group of participants. In such programmes, an individual teacher could possibly be actively involved in a programme of PD for periods of only a few weeks or months.

Over the two cluster programmes as a whole, individual teachers took an active part in professional development off and on for anywhere between two months and three years. Under one third of teachers in the first cohort and over one third in the second took part for the full three academic years, many of them lead teachers for their schools. Another third and a quarter respectively took part for around two school years, and a quarter in both cases took part for up to one school year (Table 1). Primary teachers seem to have been engaged in the programmes for significantly longer periods than secondary teachers, possibly because of a tendency in secondary schools to rotate teachers through the ICTPD programmes department by department.

Table 1: Duration of individual primary and secondary teachers' active involvement in the ICTPD programmes.

Duration of PD (months)	1999 cohort			2001 cohort		
	Primary	Secondary	Total	Primary	Secondary	Total
0–6	5.57%	17.95%	7.78%	7.63%	18.16%	10.01%
7–12	18.80%	19.87%	18.99%	16.08%	20.79%	17.43%
13–18	13.65%	16.67%	14.19%	7.36%	7.37%	7.30%
19–24	23.54%	16.03%	22.20%	14.26%	11.58%	14.10%
25–30	10.31%	8.33%	9.95%	5.54%	9.47%	6.73%
31–36	28.13%	21.15%	26.89%	41.24%	23.95%	36.31%
Number	718	156	874	1014	347	1460

Primary teachers in the second cohort also seem to have taken part for significantly longer time periods than those in the first cohort.

The importance of the third year is also apparent in terms of the frequency of teachers' classroom use of ICTs. For the 1999 cohort, statistically significant results (p<.05) were found for nine of the matched pairs of time spans, with a predominance of one-to-six month comparison items at one end of the scale and a mixture of longer time spans at the other (Table 2). It also seems clear from Table 2 that the frequency of reported classroom use was significantly less among teachers who were engaged in the programme for one to six months than among those in the programme for a year or more, and significantly greater for those in the programmes for thirty to thirty-six months than almost all other time spans.

Table 2: Confidence growth and growth in usage compared to time in programme, 1999 cohort.

Fisher's PLSD for Confidence Growth. Effect: Time span (six month intervals) Significance level: 5%					Fisher's PLSD for Usage Growth. Effect: Time span (six month intervals) Significance level: 5%				
MiP*	MD*	CD*	P-value	Sig?*	MiP*	MD*	CD*	P-value	Sig.?*
1–6,7–12	0.011	0.200	0.9137		1–6,7–12	0.113	0.229	0.3339	
1-6,13-18	0.190	0.203	0.0675		1–6,13–18	0.244	0.233	0.0398	Yes
1-6,19-24	0.186	0.183	0.0469	Yes	1–6,19–24	0.386	0.210	0.0003	Yes
1-6,25-30	0.205	0.230	0.0803		1–6,25–30	0.334	0.264	0.0130	Yes
1–6,31–36	0.596	0.171	<.0001	Yes	1–6,31–36	0.812	0.195	<.0001	Yes
1–6,7–12	0.179	0.221	0.1136		1–6,7–12	0.131	0.254	0.3092	
7–12,19–24	0.175	0.203	0.0916		7–12,19–24	0.273	0.232	0.0212	Yes
7–12,25–30	0.194	0.246	0.1218		7–12,25–30	0.221	0.282	0.1241	
7–12,31–36	0.585	0.192	<.0001	Yes	7–12,31–36	0.699	0.219	<.0001	Yes
13–18,19–24	0.004	0.206	0.9686		13–18,19–24	0.142	0.236	0.2391	
13–18,25–30	0.016	0.249	0.9023		13–18,25–30	0.090	0.285	0.5373	
13–18,31–36	0.406	0.195	<.0001	Yes	13–18,31–36	0.567	0.223	<.0001	Yes
19–24,25–30	0.020	0.232	0.8679		19–24,25–30	0.052	0.266	0.7015	
19–24,31–36	0.411	0.174	<.0001	Yes	19–24,31–36	0.426	0.199	<.0001	Yes
25–30,31–36	0.391	0.223	0.0006	Yes	25–30,31–36	0.478	0.255	0.0003	Yes
*MiP = Months in Programme, MD = Mean Difference, CD = Critical Difference, Sig? = Significance									

A similar trend is apparent for the 2001 cohort where there seems to be an even more direct correlation between length of time in the programme and significant increases in reported use of ICTs with classes (Figure 1).

Figure 1: Teachers' reported increase in frequency of use of ICTs with classes by length of time in programme, 2001–2003 cohort.

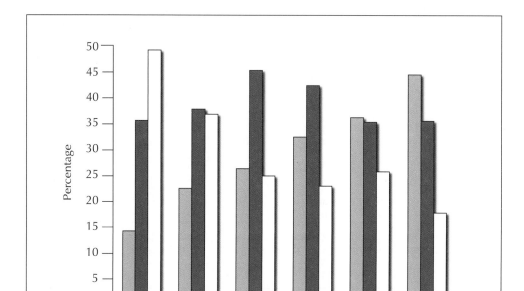

It is interesting to note, too, that in all of the case-study clusters where annual rotations of schools or teachers were planned, the one-year involvement was amended at the end of the first year and further provision was made for many or all of the participating schools to extend their time in the programme. The pressure for this came predominantly from the schools and teachers themselves, and especially from those who were taking on a lead teacher role, who felt strongly that a single year of participation was insufficient to achieve the objectives set for the programmes.

On the whole, therefore, the more successful programmes were seen to be those that maximised the amount of teacher release available to participants – and those that involved teachers in ongoing PD support over an extended period of time – this time being measured in years rather than in months or school terms.

What matters most is what you do, how you do it, and who you do it with

What you do: Programme content

In terms of programme content, the major issue seemed to concern the balance of emphasis given to personal skills, practical classroom ideas and pedagogical theory.

Compared with the other major goals of the project, which were to do with policy development and teacher use of ICTs for administration, the balance of focus on teaching and learning in each programmes' content does not seem to have been a critical factor in perceived or actual cluster effectiveness. Levels of participant satisfaction with the PD programmes were very similar, no matter whether the major focus of the PD programme was on ICT in classroom teaching and learning, or on ICT policy development and administrative applications. What does seem to have been important, however, was the balance struck by each programme within their focus on teaching and learning among three key content elements:

- *Technical skills and competencies:* The focus of these sessions was on how to operate various ICTs. The aim was to develop teachers' personal or operational skills and knowledge in relation to operating a range of ICTs that might help them in their administrative roles or, more often, that students might use in classrooms.
- *Classroom strategies:* The focus of these sessions was on 'how' and 'when' student use of ICTs might be integrated into classroom programmes. The aim was to develop practical classroom strategies for the incorporation of ICTs. This included discussion of classroom organisation techniques, as well as ideas for topics and themes that particular ICTs might be used within.
- *Pedagogical rationales:* The focus of these sessions was on the 'why' of ICT usage in classrooms. The aim was to develop teachers' broader understandings of the links and connections that might be made between ICT use on the one hand and various pedagogical or learning theories on the other.

Although the action plans of almost all clusters gave similar weights in emphasis to all three of these content elements, it was clear from the observations and the surveys that not all participants gave them similar weighting in terms of their individual goals and priorities, and nor did all clusters give them similar emphasis in the daily practice of delivering the PD programmes.

It is apparent, for example, that facilitators in both cohorts differed as a group from the teachers in terms of the focus of their goals. Whereas teachers tended to be focused in their goals on pragmatic issues related to personal skills and the practicalities of classroom use, the facilitators were often more focused on pedagogical issues. At least at the beginning of the programmes, the teachers wanted to know 'how' but most of the facilitators wanted them to explore 'why'.

In many clusters, theories of learning and teaching were prominent among the readings and activities, and in these clusters the journey for the teachers was often a shift in emphasis through the 'how' towards the 'why'. Professional readings that were used to some extent in most of the cluster programmes generated quite a few negative comments from teachers in the early stages of the programmes, and this strategy was not rated highly by them in the survey. But the totality of evidence from the clusters would suggest that this aspect of the delivery, insofar as it represents or exemplifies the particular cluster's emphasis on pedagogical theory and the connections of ICT usage to pedagogical theories, was nevertheless effective in generating teacher awareness of these issues.

As one teacher expressed this somewhat ambivalent reaction to the theoretical component of her programme: 'Whereas I thought for a start skill based learning is what I need. [And] I do, I do, I do, you know. And I thought, OK, yes, yes, the theorists, yes, we're going a bit too much into that. But no, when I look back it was actually quite a really good balance.' (Teacher interview 2001)

Our conclusion for both cohorts has been that the teachers took from the various cluster programmes according to the emphasis given. Where a strong emphasis was put on theoretical or philosophical perspectives, this often became, over time, and not without some resistance from the participants, part of the teachers' conceptual frameworks, and it often formed part of their later discourse when reflecting on the effectiveness of the programmes. Where the programme was focused primarily or exclusively on more practical issues, the simple achievement of these practical goals was the main component of teachers' reflections. Both groups, therefore, emerged from the experience relatively content that their goals had been met, but the range of ICT related issues that had been covered, and, one suspects, the depth of understandings about ICT's role in teaching and learning achieved, was rather different for the two groups. To this extent, the clusters that had a comprehensive programme which either overtly or in more subtle ways emphasised the development of participants as reflective practitioners seem to have had a particularly strong and possibly broader impact.

How you do it: ICTPD activities

If what facilitators 'delivered' in their professional development programmes seemed to have had an effect on the goals achieved, so too did the ways they delivered the context.

A question in the end-of-project surveys addressed participants' retrospective views on the particular modes or forms that PD took in their clusters. Taken alongside the interview data in which participants in the case-study clusters were asked more open-ended questions on the subject, these provide interesting indications of which particular strategies the various groups found to be most effective as PD 'events'.

Participants had some clear preferences in terms of the formats that their PD took, with teachers from both cohorts highly valuing release time, technology mentors, practicums or retreats and one-to-one tutorials. Although there was agreement among the participant groups about many of these formats, nevertheless there were a few interesting role-based differences in the value put on some particular modes of operation.

Teachers, facilitators, and principals tended to rate teacher release time, technology mentors, on-the-spot support, and to a lesser extent workshops/seminars as 'effective' modes of PD delivery used in the clusters. All three groups also tended not to rate workplace visits and listservs as highly. On one or two delivery modes, there were significant differences among the participant groups. Principals and facilitators, for example, were much more likely to rate formal conferences highly than were teachers; principals were more likely to rate school visits higher than either teachers or facilitators, and facilitators tended to see professional readings more favourably, or should we say, less unfavourably, than teachers or principals. The modes of delivery getting most 'effective' or 'very effective' ratings are shown in Table 3.

Table 3: Effective or very effective professional development programmes rated by delivery mode.

Teachers 1999 cohort		Facilitators 1999 cohort		Principals 1999 cohort	
Release time	80%	Conferences	90%	On-the-spot support	74%
Technology mentors	75%	Tutorials	89%	Technology mentors	70%
Practicums	71%	Release time	81%	Lead teachers	69%
Tutorials	71%	On-the-spot support	79%		

When the selections from the surveys are put alongside the interview data, it becomes clear that the modes of delivery most positively regarded by participants were those that involved practical and relevant skills and classroom ideas, those that maximised the time available for teachers to come to grips with the content involved, and those that combined this substantial 'time out' with ongoing access to facilitators or mentors within their schools or classrooms.

Moreover, within those modes of delivery, teachers also had decided preferences about how they liked to be grouped. The great majority of teachers found working in one-to-one situations, either with an ICT facilitator or with a colleague, to have been the most effective method for learning new skills. Ninety-four percent and 81 per cent of teachers respectively reported these one-to-one situations as having been either 'effective' or 'very effective'. The great majority (73 per cent) also felt that working in a large group (laboratory) situation was 'not effective' or only partially effective' in developing skills. As a group, teachers were fairly neutral about working on their own with written support materials for this, although secondary teachers were more likely to report this as being effective than primary teachers.

Most teachers found that working in groups with similar skills levels was more effective than working in mixed-skill level groups. They also reported that working with colleagues from their own school was more effective than working with colleagues from other schools, although this runs somewhat counter to the benefits of cross-school sharing so frequently reported in responses to other survey questions and in the interviews. Primary teachers tended to be more favourably disposed towards inter-school groups for skill development than secondary teachers.

A similar ambivalence seemed to exist about the effectiveness of working across subjects, syndicates or sectors. The majority of teachers who had worked in that way felt that working in their own departments or syndicates felt that was a very effective strategy, but those who had experienced working across syndicates or departments also reported this strategy as effective. They also seemed undecided or ambivalent about the relative effectiveness of working across sectors, with about half of the teachers who had been in such groups reporting that skill-development groups with both secondary and primary teachers in them had been 'effective' or 'very effective'.

Who you do it with:

'This cluster is strong and the schools mutually supportive. It has been very effective in areas other than ICT. The principals and BOTs are very keen to develop ICT use across the cluster. There are some very able and enthusiastic teachers who are now helping others in the cluster. We have high expectations about future possibilities. The Lead Teacher for the ICT cluster has been a strong leader with vision and innovative flair.' (Principal survey, 2001)

'Facilitators were supportive and caring.' 'A tremendous set of leaders who were inspiring and encouraging.'(Teacher survey, 2001)

'How principals could imagine that they could be divorced from [a commitment to ICTPD] by delegation, or divorced from it by inaction ... defies logic really.' (National Coordinator, 2002)

Teaching is at heart an ethical activity. It is an attempt to do good for others. And so it is, too, with teacher professional development. Its goal is to improve teaching practice, where that practice can be improved. It is to increase teachers' understandings of the social practice we call teaching, and thereby to enhance that fundamentally ethical purpose of teaching, which is to foster learning in others. Teaching, therefore – either the teaching of children in a classroom or the 'teaching of other teachers' in PD contexts – is about people interacting with other people. So, it should hardly be surprising, though it is often ignored in the research literature, to find that the key factors in providing effective PD programmes were felt by participants to revolve more than anything else around issues related to the human and professional qualities of the people involved and the nature of the social relationships established during the programme.

In other forums (Ham, 2001) I have described effective PD in the ICTPD clusters as being determined by the 'first four ships' of good clustering practice. By this characterisation, effective ICTPD clusters are those that take full and appropriate account of:

- *Leadership:* The active involvement and support of school principals and middle management in schools is essential if the PD is to have across-school effects. The abilities and expertise of the facilitators, as professional developers of their colleagues more than as technical experts or even as exemplary teachers, is an equally fundamental aspect of leadership.

- *Ownership:* How 'democratic' is the PD process? What is done in the PD to ensure that participating teachers have a say in the issues being addressed, the ways it is conducted and the methods of evaluation? Does an element of compulsion help or hinder? Do participants have a good rationale for 'why' they should (or should not) use IT, beyond just 'how'? Do teachers get what they want or what they need? And how does this change over time? Are they in the programme long enough to 'own' it?
- *Fellowship:* This relates to both the extent to which the PD addresses the affective domain and confidence (as opposed to competence) concerns of participants, as well as the extent to which opportunities are built into the structure for teachers to learn off each other, to discuss issues, to share their experiences with colleagues from other classes, schools and so on. Does the PD process build a genuine sense of community?
* *Partnership.* The extent to which the PD is well resourced outside the core Ministry funding. It includes the more general meaning of sponsorship that revolves around supplementary support in money and kind from outside agencies, businesses community groups, and so on, other than the schools themselves.

It is interesting to reflect that all four of these characteristics are commentaries on the human or interpersonal aspects of the process of PD, rather than its organisational form or even its content.

This is in keeping with some other recent research, which indicates that the things most valued by New Zealand teachers in PD are as much, if not more, to do with the professional and personal qualities of the person delivering it as with the message he or she is delivering. Davey (2000) concludes from her national survey of English teachers' views on effective PD that a human, social or interpersonal component was 'crucial in most teachers' descriptions and conceptions of "effective professional development" ' (pp. 23–4). High-quality presenters/facilitators, who were experienced, enthusiastic, highly skilled, empowering, stimulating or motivating in their facilitation and able to generate a strong sense of 'professional fellowship', were described as central to such conceptions. Our experience with the ICTPD clusters would tend to support these conclusions.

Affective descriptors such as enthusiasm, ownership, energy, innovation, flair and commitment were a very noticeable aspect of the discourse of the teachers, and even more so the facilitators in the programmes. They highlight the importance, in participants' eyes, of PD as a series of socio-professional and developmental interactions, and not merely as an organisational system or a means of information delivery. In such situations, an individual facilitator's or principal's experience, professional credibility, leadership abilities and personal mana all become important factors in determining the effectiveness of the programme. 'Who' was conducting, monitoring or managing a programme, their relative positions in the formal and informal professional hierarchies operating in the clusters and their levels of personal commitment to it seem to have been as important in determining the overall impact of a programme as 'what' was covered in it or 'how' it was organised.

In keeping with Davey's findings, all of the principals and many of the teachers interviewed at the end of their project mentioned the quality of their facilitator as a key factor in determining what worked and what did not in their cluster programme. Indeed, from some we got the impression they felt the combination of a quality facilitator and sufficiently strong leadership and buy-in from the principals could overcome most other limitations that might be present in the structure, content or mechanisms of a PD programme. The desirable qualities most valued in facilitators reflected those emerging from the study cited above: experience and sector-credibility, organisational skills, subject knowledge, enthusiasm, flair and emotional supportiveness. Comments along these lines included:

'Facilitators were supportive and caring.'

'High-quality information.'

'Excellent organisation.'

'Good flow of information.'

'The superb organisation of the project coordinators. They have put in place systems and structures which have led to accountability, have encouraged and enthused the staff participating. They have had the required expertise and have dealt professionally and encouragingly with staff to reach outcomes.'

'The expertise and enthusiasm of the facilitator has been instrumental in our developments.'

'The lead teacher for the ICT cluster has been a strong leader with vision and innovative flair.'

'A tremendous set of leaders who were inspiring and encouraging.'

As for the facilitators themselves, heavy workloads and very high self-expectation seemed to be a feature of the facilitator role, and all of the facilitators put a great deal of energy into coming to grips with what was for many of them an unaccustomed role and job description. We did note, though, that the potentially critical factor in terms of their sense of their own effectiveness seems to have been whether or not they were full-time or part-time in the role. The end-of-project survey and interviews with facilitators indicate that part-time facilitators tended to be less convinced of their effectiveness, or at least felt that they had more difficulty in being effective, than facilitators. This seems to have been especially the case where a part-time employed for the other part of their job as the ICT coordinator/teacher which was often the case in secondary schools. The conflicting time by the delivery of the ICTPD programme on the one hand and the requirement to be available for technical support or teaching within their own schools was mentioned to some degree or another by all of the facilitators who were in this situation. As one put it: 'point-five isn't really point-five – it's a lot more. So both point-five positions taken together come to equal more than full time employment ... being part-time is too frenetic, especially given the pay.' All of the full-time facilitators who responded to the 2001 end-of-project questionnaire said that their overall expectations of the programme had been 'fully met' or 'exceeded'. By contrast, only ten out of the eighteen part-time facilitators who responded expressed the same view.

If the credibility, knowledge, personality and leadership skills of the facilitators were important to the effective running of the clusters, so too was the extent of active leadership, support and commitment shown by the principals and senior managers of the participating schools. Although often not participants in the professional events themselves, their active support of the cluster concept, of the particular programme of PD in place and of the facilitator as the central figure in the process was nevertheless seen as an extremely influential factor.

Facilitators in particular spoke of the importance of promoting 'greater understanding by people at management-team level' in the ICTPD programmes, of having principals 'accepting responsibility to drive IT', of 'targeting principals as the key players', and of generally building a culture of 'valuing, rather than jockeying for position' among cluster school leaders. They spoke, too, of the effectiveness of 'regular get-togethers for all cluster principals', of appointing the school principals to any cluster management team, and of sharing funds transparently and equitably among participating schools as mechanisms for generating commitment and buy-in from the principals. There was an implicit, and sometimes explicit, assumption in the interviews that to be really effective such support had to be active rather than passive, or delegated.

Confidence, competence and classroom usage

Perhaps the final comment to be made about the 'human factors' in ICTPD relates to the importance of the role of confidence-building in the process, as implied perhaps particularly in the notion of fellowship above. PD, like classroom teaching, is as much about the affective as it is about the cognitive or behavioural domains, and this seems especially true of PD in ICT. In this respect, one of the most significant findings of the studies seems to be the apparent enduring importance of building confidence – and not just competence – in teachers about ICTs and their educative applications.

It has been noted already that facilitators emphasised the importance of the affective domain when evaluating their role in the PD process. In surveys and interviews alike, they often highlighted their perception that teachers feel less comfortable with ICT than they do on the more familiar grounds of specific curriculum areas, or even teaching and learning in general.

Teachers' stated PD goals were predominantly about the acquisition of ICT skills, and only secondarily about learning practical ideas for classroom uses. But their confidence levels on entering the programme show the opposite tendency, whereby they were much less confident about using ICTs with classes than they were about their own personal use. The interplay of confidence, competence and other factors, such as teaching experience, length of time in the PD programme, gender, and so forth, is a very complex one, but there are some indications from the cluster studies that suggest that the affective domain variable of confidence may have a greater role to play in determining the extent of classroom usage of ICTs than, for example, the behavioural domain variable of technical facility or skill level.

Both end-of-project surveys, for example, show teachers in all clusters reporting that their ICTPD programmes had led to a substantial growth in both their professional confidence about, *and* in their classroom usage of, ICTs. In the twenty-three clusters project, too, a one sample t-test with a hypothesised mean of 0 (no growth) on survey questions about the extent of teachers' growth in confidence and in classroom use both gave highly significant results ($p < .0001$). Moreover, there were also statistically significant correlations between the length of time the teachers were actively engaged in the PD programmes and their confidence, between the length of time they were engaged in the PD programme and their classroom usage, and between their growth in confidence and their growth in classroom use ($p < .0001$).

Within this general trend of a longer time in the PD programme leading to greater confidence and greater use, there are also indications that the third year of the programmes may have had a particularly strong influence in these respects. Tests for length of time on the programme (in six-month intervals) compared with confidence growth gave statistically significant results ($p < .05$) for six of the matched pairs of time spans, five of which had thirty-one to thirty-six months as their second comparison item (Table 2). The same trend is apparent in relation to the teachers' reported usage of ICTs with classes in the 1999–2001 cohort (Table 2) and the 2001–2003 cohort (Figure 1). This indicates strongly that both the growth in teachers' confidence about ICTs and their actual use of ICTs in classrooms were significantly greater among those who were in the programme for the full three years than for those who were part of it for shorter periods. By contrast, reported growths in teachers' ICT skills over the period of the programmes were not apparently determined by, or correlated with, school sector, gender or the length of time teachers were in the PD programme, and there is little evidence of a direct correlation between increased skill levels and increased classroom usage.

There are strong indications from both cohorts that have so far completed their three-year programmes, that the longer the PD, the greater the growth in teachers' confidence about classroom use of ICT; and that the greater teachers' confidence is about classroom use, the more often it happened. Logic would suggest a clear connection between growing confidence and actual classroom use. The potential importance of the finding in this case is the contrast between the clear relationships among usage, confidence and length of time in a programme, on the one hand, and the relative *lack* of connection between them and increased skill levels, on the other. In this respect, the studies tend to support the findings of another recent study, which compared the *entry-point* skills and usage levels of all four cohorts of teachers in ICTPD programmes. Ham and Graham (2004) found teachers' *entry-point* ICT skills increased over time but found relatively little change in the extent of entry-point classroom usage of ICTs. Therefore, it appears that the two variables most closely correlated with increased classroom usage in the clusters are length of time in the programme and levels of teacher confidence, rather than levels of teacher skill. If true, this has significant implications for the design of PD programmes in ICT, especially with regard to the relative attention given or not given to the development of teachers' competence or skill levels.

Conclusion

Thus, the interplay among the various variables that might combine to constitute an 'effective' PD programme in ICT is complex and for the most part irreducible to a single formula of best practice. In this respect, the national evaluations of the ICTPD programmes seem to confirm the views of those who advocate holistic and longitudinal rather than reductionist and short-term modes of evaluating PD programmes in ICT. They also add weight to an apparently growing consensus that the list of essential components should include more than just an analysis of a programme's organisational form and structure. We need to get beyond glib statements about PD being more effective because PD events are held away from, or within, a school, or are run by insider rather than outsider experts, or are individually or 'whole-school' focused, or involve the use of particular pedagogical strategies, such as skills workshops or conferences, and so on. In particular, it seems, such a list of essential components needs to include a thorough review of the content of the programme as well as an analysis of both the interactional and, perhaps even more important, interpersonal dynamics of the process. We may not yet have a recipe for effective ICTPD, but we are moving closer to an agreed list of ingredients.

Acknowledgement

The research reported in this chapter was funded through the Contract Research Division of the New Zealand Ministry of Education. It was conducted by a team of researchers from the Christchurch College of Education, the University of Canterbury and Ultralab South.

References

APEC (2004). *A Summit on education reform in the APEC region: Striking balance, sharing practice from east and west.* Unpublished Conference Proceedings and Agenda. Beijing: APEC.

Baldwin, B. & Sinclair A. (1994) *The status of technology usage in Southeastern Louisiana and the impediments to technology usage.* Paper presented at Annual Meeting of the Mid-South Educational Research Association, Nashville, TN, November, 1994.

Bolam, R. (1997*). The continuing professional development of teachers.* Swansea: GTC England and Wales Trust.

Brown, M. (1993). The School development process: A model for introducing information technology (IT). *Computers in New Zealand Schools*, 5(3), 23–5.

Cuban, L. (1993). Computers meet classroom: Classroom wins. *Teachers College Record*, 95(2), 185–210.

Cuban, L. (2003). *Oversold and underused computers in the classroom.* Cambridge, MA: Harvard University Press.

Davey, R. (2001). *Virtual PD or Real PD? New Zealand teachers' use of websites for professional development*. Paper presented at the British Educational Research Association (BERA) Conference, Leeds, September 13–15 2001.

Faseyitan, S. *et al.* (1996). An inservice model for enhancing faculty computer self-efficacy. *British Journal of Educational Technology*, *27*(3), 214–26.

Fullan, M. (1996). Professional culture and educational change. *School Psychology Review. 25(4), 496–501.*

Ham, V. (2003). *Preliminary results from the evaluation of the 23 Clusters ICTPD Programme*. Paper presented to the Learning@School Conference, Te Papa Wellington, February 2003.

Ham, V. & Graham F. (2004). *National trends in teacher participation in ICTPD Cluster Programmes, 1999–2003: Results from the baseline surveys*. Unpublished Research Report for the Ministry of Education, March 2004.

Ham, V. *et al.* (2002). *What Makes for Effective Professional Development in ICT?* Wellington: Ministry of Education.

Ham, V. *et al.* (2004). *A pre-post survey study of participants from the 28 clusters, 2001–2003. Unpublished research report for the Ministry of Education.*

ITAG (1998). *ImpacT 2001. Strategies for learning with information technology in schools.* A submission to the New Zealand government by the Minister for Information Technology's Information Technology Advisory Group.

Johnson, R. (1995). Computers and learning in primary schools: A case study in teacher development. In Tinsley, J.D. & van Weert, T.J. *World Conference on Computers in Education VI, Proceedings of WCCE '95.* London: Chapman & Hall.

Jones, R. (1997). From mountain rescuer to alpine adventure guide, or, confessions of an IT resource teacher. *Computers in New Zealand Schools, 9*(1), 25–9.

Joyce, B. & Showers, B. (1988). *Student achievement through staff development.* London: Longman.

Kellogg, D. (1996). *Getting Rural Teachers On-Line with SLIP.* Paper presented at the International Conference on Technology and Education, New Orleans, University of Texas at Austin, March 1996.

Kerr, S.T. (1991). Lever and fulcrum: educational technology in teachers' thought and practice. *Teachers College Record, 93*(1), 114–28

Knight, P. & Albaugh, P. (1997). *Training technology mentors: A model for professional development.* Paper presented at the Annual Meeting of the American Association of Colleges for Teacher Education. Phoenix, AZ: TSI.

LeBlanc, P. (1996). Project Infusion: Teachers, training and technology. *Journal of Information Technology in Teacher Education, 5*(1-2), 25–34.

McDougall, A. and Squires, D. (1997). A framework for reviewing teacher professional development programmes in information technology. *Journal of Information Technology in Teacher Education, 6*(2), 115–26.

Ministry of Education (1998). *Interactive education: An information and communication technology (ICT) strategy for schools.* Wellington: Ministry of Education.

Ministry of Education (2001). *Digital Horizons: Learning through ICT. An ICT strategy for schools. Wellington: Ministry of Education*

Ministry of Education (2003). *Digital horizons: Learning through ICT. An ICT strategy for schools. Revised. Wellington: Ministry of Education.*

Norton, P. & Sprague, D. (1996). *Changing teachers – changing schools: assessing a graduate program in technology education. Journal of Information Technology in Teacher Education, 5,* (1-2), 93–105.

Office for Standards in Education [Ofsted] (1994). *Information technology in school: The impact of the Information Technology in Schools Initiative 1990–1993.* London: Her Majesty's Stationary Office.

Owen, M. (1992). A teacher-centred model of development in the educational use of computers. *Journal of Information Technology in Teacher Education, 1(1), 127–37.*

Pelgrum, W.J & Anderson, R.E (1999). *ICT and the emerging paradigm for life long learning. Amsterdam: International Association for the Evaluation of Educational Achievement.*

Persichitte, K. & Bauer, J. (1996). *Diffusion of computer-based technologies: getting the best start. Journal of Information Technology in Teacher Education*, 3(2), 35–41.

Rhodes, V. & Cox, M. (1990). Current practice and policies for using computers in primary schools: Implications for training. *ESRC Occasional Paper InTER/15/90*, Lancaster: ESRC.

Ridgway, J. & Passey, D. (1991). *Effective in-service education for Teachers in Information Technology STAC Project.* University of Lancaster, Coventry: NCET.

Sallis, P. (1990). *Report of the Consultative Committee on Information Technology in the School Curriculum.* Wellington: Ministry of Education.

Somekh, B. (1997) Classroom Investigations. Exploring and evaluating how IT can support learning. In Somekh B. & Davis N. (eds) (1997). *Using information technology effectively in teaching and learning.* London & New York: Routledge.

Somekh, B. & Davis, N. (1997). Getting teachers started with IT and transferable skills. In Somekh, B. & Davis, N. (eds), *Using information technology effectively in teaching and learning.* London & New York. Routledge.

Stanley, D. (1995). Teacher professional development for information technology in the school curriculum. *Computers in New Zealand Schools, 7*(3), 3–7.

Watson, S. (ed.) (1993). *The ImpacT report: An evaluation of the impact of information technology on children's achievements in primary and secondary schools.* London: King's College Centre for Education Studies.

Wiburg, K. (1994). Integrating technologies into schools: Why has it been so slow? *The Computing Teacher, 1*, 6–8.

ICT Leadership: The Role of the ICT Coordinator

5

Kwok-Wing Lai and Keryn Pratt

Since 2000, a team of researchers from the Faculty of Education at the University of Otago has been investigating the use of ICT in secondary schools in the Otago region. From this work, and from the literature, a number of factors have been identified as affecting the use of ICT by staff and students in schools, and in particular the level of integration of ICT into the teaching and learning process. One issue that has been identified through this research is the role of the ICT coordinator in these schools, and the importance of this role in facilitating the effective use and integration of ICT. This chapter explores factors affecting integration levels, and particularly the impact of leadership in this area. We then look at the current roles of the ICT coordinators in Otago schools, and how they are placed to provide this leadership.

Factors Affecting the Integration of ICT

Our own research, and that of others, has highlighted the impact that access to ICT has in terms of its subsequent use and integration into teaching and learning (Lai & Pratt, 2002; Lai *et al.*, 2001; Lai, Pratt & Trewern, 2002; Murray & Campbell, 2000; Sheingold & Hadley, 1990). For example, Smerdon *et al.*, (2000) and Becker (1999) both reported that teachers were more likely to use ICT if it were readily available in their classrooms. A lack of access was identified by Smerdon *et al.* (2000) as being one of the two most significant barriers for technology use of secondary teachers in the United States (US), while Becker (1999) found that the level of classroom Internet connectivity was the most important variable in predicting U.S. teachers' Internet use with students. Lawton (1994) also emphasised that the equipment must not only be available but also be working if it were to be integrated into teaching and learning.

In addition to being able to access ICT, teachers also need to have the necessary skills and confidence to use it, and to integrate it into the teaching and learning process (Becta, 2000; CEO Forum, 1999; Hargreaves, 1994; Lai & Pratt, 2002; Lai *et al.*, 2001), with Murray and Campbell (2000) finding that a lack of teacher training and limited levels of computer skills were obstacles to the effective use of ICT. The impact of teachers' skills on usage can be seen in Becker's (1999) finding that one of the predictors of US teachers' Internet use with students was their level of overall computer expertise. Smerdon *et al.* (2000) also reported that teachers who felt better prepared to use ICT in the classroom were more likely to do so, while Falloon (n.d.) found that teachers' skill and training with computers contributed to and supported exemplary computer use in the primary classrooms.

Another factor that has been found to affect teachers' use of ICT was their attitudes towards technology and beliefs regarding its value in education (Lai *et al.*, 2001). Becker (1999) reported that teachers' pedagogical beliefs was a predictor of their Internet use with students, and Falloon (n.d.) found that the attitudes teachers had towards technology, their beliefs regarding the value of using computers with children and teachers' pedagogical beliefs contributed to and supported exemplary computer use in the primary classroom.

The importance of giving teachers time to learn to use ICT, to implement it in their teaching practice and to reflect on the effectiveness of this implementation was emphasised by a number of researchers, including Cook (1997), Ang (1998), Glennan & Melmad (1996) and the National Education Association [NEA], (1999–2000). Smerdon *et al.* (2000) found that a lack of time to learn how to use computers was identified by U.S. secondary teachers as the second of the two most significant barriers, and in New Zealand Murray & Campbell (2000) reported that time, workload and curriculum pressures were seen as obstacles to using ICT effectively.

Teachers also need support if they are to use ICT effectively, with Sheingold and Hadley (1990) noting that teachers needed support in learning and planning how to use the technology, and Ringstaff *et al.* (1995) and Means & Olson (1995) both commenting on the importance of having readily available technical support if teachers are to be encouraged to use ICT.

Many of these issues are related to the process of undertaking change in general. Fullan (1993) noted that change processes, particularly those involving teaching and learning, were particularly complex, and commented that: 'The hardest core to crack is the learning core – changes in instructional practices and in the culture of teaching toward greater collaborative relationships among students, teachers and other potential partners...Changing formal structures is not the same as changing norms, habits, skills and beliefs' (p. 49). Fullan identified a number of other factors that made this process of change more complex than some others, including the need for changes in the existing power structure, and the involvement of large numbers of people, such as teachers, management, students and their families. The complex nature of change of this type can be seen in the factors that have already been identified as affecting the integration of ICT into teaching and learning, including the new or increased need for time, support and professional development, as well as the effect of attitudes and beliefs on the level of integration. Sheingold & Hadley (1990) found that in order for ICT to be more widely used, there needed to be both a school structure and culture capable of encouraging its use. Ringstaff *et al.* (1995) also noted the importance of support within the school in their work with the Apple Classrooms of Tomorrow (ACOT) Teacher Development Center. They found that, of those involved in their projects, teachers who had the support of the principal were more likely to use technology with their classes than those without this support. This support is considered necessary, as integrating technology into everyday lessons may require adjustments to equipment purchasing, and to school scheduling, to allow teachers time to plan and collaborate on complex projects.

Leadership in ICT

Levels of leadership have been shown to affect the level of ICT use and integration in schools (e.g. Becta, 2002; Schiff & Solmon, 1999). In their evaluation of the first year of the California Digital High School project, Schiff & Solmon (1999) found that schools needed to have strong leadership and a well-designed technology plan to be successful users of ICT. Becta's (2002) report on the effect of ICT on educational standards also identified leadership as a contributing factor in determining whether or not British primary schools would make good use of ICT, and they listed five factors that needed to be present for schools to do this: ICT resources; ICT teaching; ICT leadership; general teaching; and general school leadership. They noted that: 'Although ICT opportunities are typically provided by the classroom teachers, the quality of leadership and management of ICT in a school is crucial to the provision of good ICT learning opportunities. As the quality of ICT leadership improves, so does the percentage of schools providing good quality ICT learning opportunities' (pp. 20–1).

It is perhaps not surprising that the research of Becta (2002) and Schiff & Solmon (1999) found that good leadership was necessary for schools to make effective use of ICT. A number of researchers (e.g. Fullan, 1992, 2001; Hopkins, *et al.* 1994; Lai, 1999) have identified leadership as being of particular importance when schools are undergoing change, including the integration of ICT into teaching and learning. When talking about the experience of New Zealand schools, Lai (1999) commented that: 'The aim is to transfer the culture of the schools to one in which ICT is seen as an agent of changing relationships between students and teachers and between learners and knowledge construction. Leadership is essential in generating a vision for the school regarding the use of ICT' (pp. 18–9).

Fullan (1992, 2001) has also highlighted the importance of leadership for schools undergoing change, and in particular the role of the principal as leader. He noted 'it should be absolutely clear that school improvement is an organizational phenomenon and therefore the principal, as leader, is the key for better or for worse' (Fullan, 2001, p. 146). The importance of the principal as the leader of initiatives, such as the integration of ICT, has been recognised in New Zealand, with the Ministry of Education's Principals First workshops being an example. These workshops were 'for principals to develop leaderships skills in planning school implementation of ICT' (Education Review Office, 2001, p. 3).

Although recognising the importance of the principal as a key change agent for innovations in the school (Fullan, 1991), the notion of shared leadership at the school level has also been suggested as appropriate for the implementation of innovations (e.g. Fullan, 1992; Hopkins *et al.* 1994). In his 1992 book on school improvement, Fullan noted that although the support of principals for innovations is critical, they 'often depend on assistance from a "second change agent" in the school' (Hall & Hord, 1987, cited in Fullan, 1992). Hopkins *et al.* (1994) also support the notion of school leadership coming from other than the principal. They believed there was: 'Growing recognition that a school which looks to the headteacher as the single source of direction and inspiration is severely constrained. It is dependent on a single individual's supply of intellectual, emotional and physical energy – it is restricted by a single imagination' (p. 155). In their work on a school improvement programme with schools in East Anglia, North London and Yorkshire, they recommended schools appoint at least two coordinators, at least one of whom is part of the school's senior management team.

The coordinators' role involved taking charge of the day-to-day project activities, and facilitating, enabling and encouraging others to fulfil the project's aims. Hopkins *et al.* (1994) recognised that appointing a coordinator means 'there is likely to be a significant change in the degree of delegation of power from senior management to staff' (p. 174). They also noted that the principal still had an important role to play:

> The effective functioning of the coordinator in this situation would appear to be crucially dependent upon the quality of delegation from the head, who needs to create a climate in which staff as a whole are able to respect and relate to the coordinator. It appears that this can best be achieved when:
>
> - there is positive support for, and effective definition of the limits of, the responsibility of both the coordinator and any groups which may be established;
> - all relevant matters are communicated to or channelled through the coordinator;
> - the coordinator is able to feel supported by the whole of the senior management team;
> - the coordinator's role is clearly understood by all of the staff;
> - the staff are able to relate to him or her on a personal level (p. 174).

In terms of innovation relating to ICT, the school ICT coordinator (also known as the technology or computer coordinator) would appear to be the most logical alternative to the principal to act as leader in this area.

The Role of the ICT Coordinator

Nearly two decades ago, the first Electronic Learning Computer Coordinator survey was conducted (see Strudler *et al.* 2001). This was followed with further surveys in 1987 and 1989 (Bruder, 1990). For the purposes of these surveys, computer coordinators were defined as 'an individual involved in planning for, purchasing, or maintaining educational technology' (Bruder, 1990, p. 24), with the 1989 results based on 666 responses from the 2000 surveys distributed to US schools. The computer coordinators involved in this survey were employed at either a district or school level, and were 'primarily administrators' (p. 29).

In the 1989 survey, the majority of school-level computer coordinators were male, experienced teachers who were currently teaching in addition to carrying out part-time duties as computer coordinators. Most of them received no compensation or salary for their additional coordinator role. The school-level computer coordinators spent their time on a number of tasks, with most time being spent on:

- teaching computer related classes or supervising student computer use
- planning or implementing computer related programmes
- training teachers
- maintaining computer equipment (adapted from Bruder, 1990).

In her discussion of the findings of the 1989 survey compared with the previous two, Bruder (1990) noted that: 'At both the school and district levels, coordinators have experienced a dramatic shift away from traditional pedagogical duties – such as teaching students and other teachers – to more administrative tasks, such as evaluating and recommending hardware and software for purchase' (p. 25).

In a smaller-scale, but more detailed investigation of the role of technology coordinators in three U.S. primary schools at around the same time, Strudler & Gall (1988) found that coordinators had a number of responsibilities, similar to those described by Bruder (1990), including training, supporting and energising teachers, providing technical support and organising the school's instructional computing programme.

Moursund (1992) summarised the duties of ICT coordinators, identifying four general categories: working as a computer facilities manager; working with school administrators and district-level educators; working with teachers; and working with students.

He also identified thirteen specific duties of ICT coordinators, including: technical support; short- and long-term planning of instructional use of ICT; helping teachers to develop curriculum materials and lesson plans; providing professional development for teachers; managing hardware, software and resource budgets; supporting students; evaluating ICT programmes; and professional learning. A study exploring an ICT initiative in British schools (Somekh *et al.* 2001) found that the ICT coordinators in many of the schools were also responsible for a number of areas, including professional development, students' skill levels, the development of infrastructure and the use of ICT to support teaching and learning.

In his case study, Marcovitz (2000) reported his observations of a part-time ICT coordinator's work during one day at a US elementary school. He identified a number of roles fulfilled by the coordinator, namely 'technician, trainer, curriculum resource, and policy maker' (p. 259), and observed three different ways in which the ICT coordinator fulfilled these roles. First, Marcovitz observed the coordinator providing 'support by walking around' (p. 260): the ICT coordinator spent a lot of time walking around the school as part of his duties, and Marcovitz observed him providing a lot of support simply through 'being in the right place at the right time' (p. 260), and gaining a lot of information about ICT in the school, which aided him in determining levels of need and policy issues. The second way in which the ICT coordinator in this school fulfilled his role was through 'nuts-and-bolts' activities, or activities involving technical, rather than educational or administrative issues. Finally, although no formal policy decisions were made during the time Marcovitz was observing him, throughout the day the ICT coordinator had informal discussions with staff, 'finding out their needs, fixing their problems, and discussing their policy concerns' (p. 268).

The technology coordinators (later called educational computing strategists or ECS) in a Las Vegas school district were found by Strudler *et al.* (2001) to be similar to those described previously, both in terms of their demographics and their responsibilities. Once again the majority of coordinators were male teachers, and had a wide range of responsibilities, including 'providing staff development and support, performing basic maintenance of hardware and software, and leading technology planning and coordination' (p. 7).

Ronnkvist *et al.* (2000) explored technology support in a broader sense, with the provision of facilities and incentives for teachers being seen as a necessary part of support as well as personnel to provide assistance, guidance and professional development. They also made a differentiation about the content of the support, seeing it as either instructional or technical. The majority of the schools who responded to their survey had a technology coordinator; however, only around one-fifth of these were full-time, with high schools more likely to have a full-time coordinator than middle or elementary schools. The part-time coordinators identified themselves as also being classroom teachers, network coordinators, media specialists or having another role. As in the research already described, these ICT coordinators had a number of responsibilities. As part of their duties, per week they spent an average of three to four hours providing one-to-one support, one to two hours per week on training and one hour helping teachers integrate ICT into their teaching.

Overall, it appears that ICT coordinators in schools around the world fulfil similar roles. Their key purpose is generally accepted to be the enhancement of teaching and learning through ICT, but they also tend to have a number of other ICT-related roles. Although various researchers have categorised these in various ways, the roles of ICT coordinators include technical, administrative and curriculum functions, with various tasks/areas in each of these. These can be summarised as follows:

- Technical
  — maintenance (ensuring hardware/software is working)
  — planning (of infrastructure, hardware and software needs)
- Administrative
  — managing accounts, etc
  — policies (health and safety policies, acceptable use policies, etc)
  — paperwork/budgets
- Curriculum
  — resources (identification, evaluation, distribution)
  — training
    skills (how to use hardware/software)
    integration (how to use hardware/software to enhance teaching and learning)
  — envisioning (looking to the future in terms of how to use ICT to enhance teaching and learning, and ensuring staff share this vision)

Although not all of the ICT coordinators were involved in all aspects of these duties, most appear to have, or would like to have, at least some role in these areas.

Drawing on the literature, it is apparent that the main role the coordinator has been expected to fill is the promotion of ICT in teaching and learning, with this remaining largely unchanged over a number of years. For example, research by Moallem *et al.* (1996) found that teachers in all six US middle schools surveyed expected ICT coordinators to provide them with instructional support through workshops and by demonstrating the application of software. In line with this, Reilly (1999) noted that it was important that the ICT coordinator's role was that of curriculum leader, not 'electronic janitor', and Lucock & Underwood (2001) observed that as teachers, the main role of the coordinator should be to guide ICT teaching and learning. These expectations, however, have not always been met.

The activity that has been identified in the literature as placing the greatest restriction on the ICT coordinators' ability to improve the use of ICT in teaching and learning in their school is the time they are required to spend providing technical support. The research by Moallem *et al.* (1996) found that ICT coordinators spent 75 per cent of their time on technical support, and Marcovitz (2000) found that out of fifty activities completed by the ICT coordinator in his case study on one day, thirty-seven were technical, with a number of the remaining thirteen activities including at least some technical aspects. Similarly, Strudler *et al.* (2001) found that the coordinators were spending more time than they wished on technical support, and less time on instructional or pedagogical support. This was despite the latter being identified as a primary function of the coordinators. Reilly (1999) experienced similar problems in his time as a school district technology coordinator. He identified the key issue he faced as the gap between his expectations of a role as 'curriculum leader' and the reality of being an 'electronic janitor'. British ICT coordinators faced a similar issue, with Somekh *et al.* (2001) noting that coordinating the use of ICT to support teaching and learning was usually the area that received little attention when they were needed to provide technical support.

Somekh *et al.* (2001) noted the impact that coordinators had on the uptake of ICT within the school, commenting that the coordinators' approach to ICT affected the approach of the whole school and, where their beliefs differed from that of the school's management, ICT use among teachers was generally not successful or widespread. Strudler's (1994) research also found that school-based technology coordinators, as change agents, could support teachers to overcome various obstacles encountered when using technology in their teaching and suggests that, without the support the technology coordinators provide, it is unlikely that technology would have an impact on teaching and learning.

The view expressed by Strudler (1994) that technology coordinators could act as change agents echoed those of November (1990), who believed that the early role of computer coordinators to coordinate computer courses and act as technicians was changing to become that of a 'broker/collaborator/change agent' (p. 8). The ability of ICT coordinators to act as change agents has been recognised by initiatives such as that described by Strudler *et al.* (2001). This initiative saw the provision of technology coordinator positions in schools, with one of their goals being to act as leaders in technology.

ICT Coordinators in Otago Schools

A research team from the Faculty of Education at the University of Otago investigated factors affecting the use of ICT in twenty-six Otago secondary schools in 2000, and again in 2002. At each of these times, ICT coordinators in these schools were asked to complete questionnaires, with eighteen doing so in 2000 and twenty-three in 2002. Interviews were conducted with seven in 2000 and fourteen in 2002. Over the two studies, information was received from the ICT coordinators of twenty-four of the twenty-six Otago secondary schools. This research has enabled us to explore the roles these ICT coordinators fill within their schools, and how they are placed to provide leadership in the integration of ICT within their schools (Lai, Trewern & Pratt, 2002; Lai and Pratt, 2004).

Who were these ICT coordinators?

At the twenty-four schools from which information was received, twenty-eight people had held the position of ICT coordinator in 2000 and/or 2002. In four schools, the ICT coordinator in 2002 had not been the ICT coordinator in 2000. Generally, the changes in ICT coordinator occurred due to staffing changes at the schools.

The majority of ICT coordinators were male, with only four female ICT coordinators. Three of these were at all-girls' schools, with the remaining one teaching at a co-educational school. Only one of the ICT coordinators had no teaching role, being employed as network manager and ICT coordinator in 2000; by 2002, the ICT coordinator of this school was a teacher. The ICT coordinators were generally very experienced teachers. In 2000, the teachers in this position had an average of over twenty years of teaching experience, while the 2002 ICT coordinators had an average of eighteen years of teaching experience, with a range of between two and thirty-four years.

With the exception of a network manager employed by one school in 2000, the position of ICT coordinator was an add-on responsibility. In 2000, the ICT coordinators were either deputy or assistant principals (28 per cent) or department heads of mathematics, technology, or ICT (61 per cent). In 2002, the majority of the ICT coordinators were still either senior managers or senior teachers, which was typical in New Zealand secondary schools, as evident from findings of previous studies conducted by Lai and his colleagues (Lai, 2001; Lai *et al.,* 1999). Eight of the ICT coordinators in 2002 were in senior management positions (principal, deputy principal, assistant principal), while six were also network managers in their schools. One of these ICT coordinators was also the head of department (HOD) of mathematics, and six of the ICT coordinators reported being the HOD or teacher-in-charge of ICT or technology. One ICT coordinator did not report any responsibilities other than those associated with the role as a teacher, while all teachers taught at least one class.

The ICT coordinators differed from classroom teachers in terms of their skills, their use of ICT and their training, as described below.

ICT coordinators were more highly skilled, in terms of ICT, than teachers: In 2002, ICT coordinators and teachers were asked to rate their ICT skills using the Mankato scale (http://www.bham.wednet.edu/tcomp.htm). The ICT coordinators' mean skill level on each of the areas measured by this scale (basic computer use, file management, word processing, spreadsheets, databases, graphics, email, research/information seeking/desktop publishing, video production, technology presentation and the

Internet) was significantly better than that of teachers. The ICT coordinators also had better technical skills than teachers in 2002. At this time they were significantly more likely to be able to fix both minor and major problems with ICT than teachers.

ICT coordinators used ICT more often for preparation and administration than did teachers: In both 2000 and 2002, ICT coordinators reported using email, the Internet, word processing/desktop publishing and computers in general more often for these purposes than teachers. In both surveys they were also more likely to use email to communicate with colleagues and senior management than teachers; in 2002, they were also more likely to use ICT for professional learning than teachers.

ICT coordinators were more likely to be using ICT in their teaching than teachers: Although in 2000 there was no significant difference between teachers and ICT coordinators in terms of using ICT in their teaching, or in requiring or allowing students to use technology, by 2002 the coordinators (excluding those who solely taught ICT-related courses) were using ICT in their teaching to a greater extent than teachers. In addition, at both time points, coordinators were more likely than teachers to use some aspects of ICT with their students. In 2000, these were email, spreadsheets/graphing, scanning/video/photo editing and graphics/drawing packages/clip art. By 2002, coordinators used word processing, the Web to search for information, spreadsheets/graphing, CD-Roms to gather information, presentation software, web-page design and peripherals with their students more than teachers; they were also more likely to expect their students to use word processing, the Web to search for information, email to work with other students, presentation software, and peripherals in their learning than were teachers.

ICT coordinators were more likely to be integrating ICT into their teaching than teachers: ICT integration levels were measured using Knezek and Christensen's (1999) six-stage model of technology adoption. The six stages in this model are: (1) awareness; (2) learning the process; (3) understanding and application of the process; (4) familiarity and confidence; (5) adaptation to other contexts; and (6) creative application to new context. In both 2000 and 2002, the mean stage of technology adoption of ICT coordinators was one stage higher than that of teachers.

ICT coordinators spent more time doing ICT professional development than teachers: Data collected in 2000 indicated that there was no difference in the level of ICTPD activities undertaken by teachers and coordinators. In 2002, however, although teachers and coordinators did not differ in their perceptions of the importance of ICTPD, coordinators spent more hours in professional development in terms of: development of units of work for using technology in your curriculum area; workshops for envisioning/motivating/enthusing staff to use computers; school/workplace visits; and attending conferences.

What did their role as ICT coordinator entail?

The coordinators at these schools engaged in a range of tasks over and above their normal teaching and/or school management duties, as shown by the following comments:

> My role is really to decide what new equipment will be purchased in the school, if there's any physical alteration to the buildings that need to be done. If there's any cabling that needs to be laid, if there's software to be purchased that is used by the whole school ... If there's computers that need to be upgraded teachers will come to me ... I s'pose I'm also responsible, I guess, for the long-term plans of where computing's going in the school. I also do some minor technician's work ... You know, the printer doesn't go, come and fix it; I can't find this programme; how do I do? So I do a little bit of kind of on-the-spot technician and a little bit of kind of on-the-spot professional development ... I also maintain the student, maintain this network in here on, I collect all the e-mail, I help to maintain the school's website although I don't write it myself ... and I suppose just the general, if problems crop up people will come to me and say what do I do about this (ICT coordinator, provincial school, 2000).
>
> Showing how to mail merge, clearing printer jams, where the on/off button is! Everything and anything in fact – all day long – at intervals and lunchtimes I'm frequently tempted to hide (ICT coordinator, urban school, 2002).

The tasks carried out by these coordinators varied somewhat between schools, but generally involved tasks similar to those identified in research overseas regarding the role of ICT coordinators.

Technical

All coordinators had at least some technical aspect to their role. This may have included managing networks, purchasing and maintaining equipment, fixing problems or keeping up to date with new innovations and using this information to make plans for the school in terms of ICT.

Maintenance

In 2002, most ICT coordinators (87 per cent) provided at least some technical support, either in terms of regular maintenance tasks or through fixing problems as they occur.

> It just goes, and goes and goes. There's always technical support. There's always something in a network of seventy computers which isn't going to work. So you've got to fix it, because people can't do anything (ICT coordinator, provincial school, 2002).
>
> The weekly, and recently daily, updating the virus protection (ICT coordinator, rural school, 2002).

Planning

The ICT coordinators in this study had great influence in planning and formulating plans for their schools, usually in terms of what was needed in terms of ICT to fulfil the school's vision for teaching and learning.

> I think I get a lot of choice in deciding how to get from where we are to where you envision (ICT coordinator, urban school, 2002).

> The ICT [plan] has been mostly developed by me over a period of about four years, with input from other teachers who say what they want, but there's also quite a large group of teachers who don't know what they want (ICT coordinator, urban school, 2002).

Frequently, coordinators developed this plan in conjunction with their principal or an ICT committee, and with input from other staff.

> The [principal] has always been very, very supportive. But for his role … it has always been administrative … so he hasn't had the thought or the vision to go and do what I've done, that's been left up to me to go away and dream and go and build something. And so that's what I've done (ICT coordinator, provincial school, 2002).

> What I do is I tend to liaise with the rest of the staff … I put together a proposal, just a draft proposal of next steps and we talked together, anyone who's interested in ICT in school and we had a meeting with the principal and talked about what possibilities might be to give a clear indication, and then I went back and modified the draft … I think the liaison's really important with what's needed and anecdotal information from students as well as to what they're after. The final decision rests with the Board and the principal, so all I ever do is recommend (ICT coordinator, urban school, 2002).

> I work with our IT advisor. We look at what we need to make the system work within the parameters of the money we've got. We take that to our IT committee which has a board member, principal, a couple of teachers and our advisor, myself … So it's a fairly large committee really but it works because they don't really make decisions. We take proposals to them, we explain carefully what it is that we want. I will have one-on-one meetings with the principal so that I've got an idea of what her vision is and we try and bring all those things together. The committee's job really is probably to identify things that our advisor and I have possibly overlooked (ICT coordinator, urban school, 2002).

In line with their role in formulating plans for their schools, all the ICT coordinators reported having some say in terms of the purchase and placement of computers.

Administrative

Many of the coordinators also dealt with the day-to-day administrative tasks involved in managing ICT at their schools.

Managing access

With the majority of schools (78 per cent) having networks, many ICT coordinators are responsible for the management of this, such as setting up user accounts.

Just the routine stuff where you know the kid who leaves and you take his record off the system and the new one who arrives, they are not individually very time-consuming but it's just five minutes here and ten minutes there and it's once, twice, three times a week sort of stuff, especially at this time of year we're getting a lot of leavers (ICT coordinator, provincial school, 2000).

I ended up setting up all permissions and policies for 1100 users, you know, and entering them with passwords and groupings and all that sort of thing (ICT coordinator, urban school, 2002).

Policies

The increasing use of ICT in schools has lead to a need for policies regarding their use to be developed and enforced. In many cases, these tasks were undertaken by the ICT coordinators.

You've spoken to that kid about putting their own software on the system and how they signed a form which said they weren't going to do that sort of stuff and then the next thing some other kid pops up and they've done the same thing – yeah that has become more time consuming (ICT coordinator, provincial school, 2000).

I've just drawn up a policy on health issues and an acceptable-use policy which we're just looking at at the moment (ICT coordinator, provincial school, 2002).

We have a users' agreement that each student and their parents has to sign each year and I keep all that information here ... We use blocking software ... I monitor that use from my computer in the lunch hour (ICT coordinator, urban school, 2002).

Saturday morning, I come in and I check the logs ... look for any dodgy addresses, go back through the times (ICT coordinator, rural school, 2002).

Budgets

In many cases, the ICT coordinators also had at least some role in terms of budgetary and resource allocation. The coordinators in this study were responsible for purchasing and placement of hardware (100 per cent) and software (91 per cent) and over half (56 per cent) of them also engaged in software selection and evaluation. Generally speaking, the ICT coordinators were responsible for purchasing software that was to be used throughout the school, with individual departments purchasing specialised software. The ICT coordinators were often consulted with regards to the specialised software available and its hardware requirements. Most of the principals relied on them to provide the information necessary for decisions relating to the purchase of ICT hardware, either on their own or in conjunction with a committee, 'in discussion with Principal and ICT group' (ICT coordinator, provincial school, 2002). They were 'always consulted and usually have a large say in decisions' (ICT coordinator, urban school, 2002) and they also provided advice to the boards of trustees.

I have the ultimate say in any purchases, when we purchase the hardware ... But I'm always advised by people like [the ICT coordinator], and ... people just can't go out and purchase, they have to go through him ... I always then say to [the ICT coordinator], well where does this request fit in our understanding? And he says, don't be silly, or he says, it's reasonable, or whatever (Principal, urban school, 2002).

All network administration software is my responsibility. Principal has introduced MUSAC [a school administration software]. Specialist software purchased after HODs have consulted coordinator (ICT coordinator, urban school, 2002).

In addition, coordinators had an awareness of health issues that had to be taken into account when ICT purchases were considered: 'The fancy chairs, in there, but that's because I brought them there ... Now, I want these rooms carpeted, that's going to happen, but I'm last on the list because I want anti-static carpet, I'm not prepared to tolerate anything that's nylon ... and I've just got to reduce the stress on computers and people by having the right stuff in' (ICT coordinator, provincial school, 2002).

Curriculum

Providing curriculum support was an important part of many ICT coordinators' role, with just over one third (34.8 per cent) indicating in 2002 that this was their primary role, with a further 30 per cent indicating that technical and curriculum support were of equal importance in their role. As one ICT coordinator commented, 'My main role's curriculum ... Mainly getting teachers up to speed umm, with software and using computers in general, you know, targeting particular software to curriculum needs or immediate needs at the time' (ICT coordinator, provincial school, 2002).

These ICT coordinators, like those in overseas schools, provided curriculum support in a variety of ways.

Resources

One of the ways ICT coordinators provided curriculum support was through the identification, evaluation and distribution of appropriate resources. In one school, the coordinator did this via 'anonymous' emails: 'I also set up a little helper which ... would just fire out to everybody on the staff just little sites that people might go to' (ICT coordinator, urban school, 2002).

Training

These ICT coordinators were more likely to be responsible for the provision of ICTPD than the teachers in these schools, with only 21.4 per cent of ICT coordinators never doing this, and 42.9 per cent regularly doing so.

In-house PD was provided in a variety of ways. Examples found in 2002 included 'breakfast with gifts' (provincial school), a 'buddy system' (urban school), and serving as a walking 'help desk' (provincial school); in 2000, the PD offered by ICT coordinators included in-school workshops and seminars, and mentoring, or team teaching with other teachers. Other approaches included providing short practical tasks, as the following shows: 'For email, all I did was email everybody and say, reply to this and I want you to attach the answer to the question that I've attached which was "four plus four" and if they emailed me back in any way, shape or form, an eight as an attachment, then I knew there was no need for any in-service' (ICT coordinator, urban school, 2002).

The professional development provided by the ICT coordinators can be divided into two types: skills-based and integration.

Skills: ICT coordinators provided skills-based training on a wide variety of areas. Some of the training was scheduled, while some was provided on an as needed basis.

> Last year I used to take ... a whole range of mini-courses and we, generally speaking, worked on say four or five people at a time on particular topics like how to use email, how to search on the Internet search engine and anything that people particularly wanted (ICT coordinator, provincial school, 2002).

> We're also looking at introducing Classroom Manager so that would be a, a big learning curve for everyone. I'm sort of co-ordinating that (ICT coordinator, urban school, 2002).

> Last year I ran a course for everybody to teach them how to use the email and we did a little bit on using a web browser (ICT coordinator, urban school, 2002).

> I see that as being part of my role so I'm not worried when somebody comes up to me and says, my machine's locked up or I can't find a file on the network. It's like, well, okay, let's go and find it but I make sure that they watch what I do or even get them to do it so that next time around, they can do it themselves and hopefully we break, and it does work (ICT coordinator, urban school, 2002).

Integration: Other professional development provided by coordinators was classroom-focused, aimed at helping teachers use ICT in their teaching: 'We've got a buddy system that exists in the school where there's no more than five teachers with a buddy and, we fire that up from time to time ... those smaller clusters of staff would actually say, well okay, I know how to do this but in my class now, I'm doing this, how can I do that and the buddies would kick in' (ICT coordinator, urban school, 2002).

Envisioning

The role of many of the ICT coordinators also included envisioning and leading staff. They usually had a vision of ICT use in their school and were drivers of this vision, either on their own or in conjunction with the principal and/or other teachers.

> Most probably even though we've got an ICT plan with a vision, it's most probably my vision more than the whole school's vision. So most probably I'm sort of trying to set the

scene so that it happens as best as I can with, so, part of it's sort of just figuring out where we're going as a school (ICT coordinator, rural school, 2000).

We're the people who are driving ICT in the school at the moment and I see them continuing that (ICT coordinator, provincial school, 2002).

That linking with teachers and even that sort of development of a culture of innovation ... within the school but is across schools, I think it most probably is where I actually see things potentially happening within the constraints of the ... secondary system. In fact I think that possibly we could go a step further, we may actually be able to sort of encourage change at a system level because we can show there are actually better ... ways (ICT coordinator, rural school, 2002).

Primary role

The roles of the ICT coordinators in these Otago secondary schools differed, with around one third indicating their role was primarily technical (34.8 per cent), the same number indicating that their role was primarily curriculum (34.8 per cent) and the remaining coordinators indicating their role was a fairly equal combination of both (30.4 per cent). This is slightly different from overseas findings, where it appeared that the main focus was the promotion of ICT in teaching and learning. This may, however, be due to the way that needs for technical support tend to take precedence over curriculum support (Somekh *et al.*, 2001). It may be that for some of these ICT coordinators, their role was primarily technical through necessity, rather than desire. A number of the coordinators talked about the competing needs of technical and curriculum needs, and of the need for teachers to be supported by fellow teachers.

But that's how I see my role, is we talk about the teaching programme, because you have two people sharing ideas, you can develop something. I mean neither of us would come up with this alone but together ... And that's not a job for a technician. That is a job for an educator (ICT coordinator, urban school, 2002).

Many, many teachers have come to me and I've helped them do different things but again, it's a time thing and, it's usually when they want to know how to do it, and that's nice and it's fair enough and, yeah, I'd consider that to be my job rather than all the technical things that I end up doing (ICT coordinator, urban school, 2002)

The need for more curriculum support was identified by teachers in these schools: more than 70 per cent of the teachers reported having no or very limited help with using ICT in their particular curriculum area (Lai, Pratt & Trewern, 2002). Teachers at these schools were asking for: 'Ideas about new stuff, ways of using materials. Evidence about best practice for use of technologies' (teacher, provincial school, 2002), and 'Someone available with vision re teaching possibilities and use of ICT' (teacher, urban school, 2002).

Curriculum support was not the only area that was perceived as lacking, however, with teachers also feeling that the importance of technical support was not only being overlooked, but that too much was asked of teachers with responsibility for this (Lai, Pratt & Trewern, 2002).

The roles filled by the ICT coordinators in these Otago schools appear very similar to those of teachers in similar positions overseas and, like their overseas counterparts, ICT coordinators here would appear to be in a position to provide ICT leadership, and in many cases are already doing so. What then, prevents their effectively leading as changes within their schools?

How are They Placed in Terms of Providing Leadership?

Hopkins *et al.* (1994) identified a number of factors that were important if teachers other than the principal were to act as leaders or coordinators within schools. These factors, discussed previously, included aspects related to the potential leader, such as an ability to relate well with all staff, as well as to wider issues. They felt that the coordinator's role and responsibilities must be clearly understood, recognised and supported by all staff, with senior management promoting this. The question then is: Whether this climate of recognition and support exist in these schools? Generally, it would appear that although these ICT coordinators played some role in leading ICT within their schools, their leadership was generally not recognised and they were not given the support necessary for it to be effective. According to these coordinators, their biggest problem was the time that was needed to fill the coordinator role, and the effect this had on aspects of their work. Overall, in 2002, 59 per cent received a time allocation or some other recognition for their work as the ICT coordinator for their school. Also in 2002, 60 per cent of the ICT coordinators were frequently asked to deliver professional development in their schools, but only 13 per cent were frequently given time to deliver it.

Recognition and Support

One school recognised the need to provide its ICT coordinator with time, by allowing the coordinator to teach the minimum allowable hours to be considered a teacher, while devoting the rest of his time to his coordinator role. Most schools, however, were not able to provide this sort of support, resulting in coordinators spending significant portions of their own time filling this role:

I get two hours a week and I get, a certain PR salary as well, but two hours is very rarely enough (ICT coordinator, urban school, 2002).

Yeah. I still have this issue of time, for people like myself, that it did, I won't say it takes every waking hour, but it certainly takes huge amounts of my time. Way, way out of proportion to the, if you like, the seniority I've got in the system because of it ... I in theory run a small department, in the school, as well as the network. And both are supposed of equal ... Equal value. But the network takes, oh, I don't ... over a year, hundreds and hundreds of hours more than does running the department (ICT coordinator, provincial school, 2002).

I've got two hours allowed. I spend between twenty and thirty hours every week over and above my teaching commitment. Yeah, so I work a sixty, seventy hour week and that's why I said I can't keep it up (ICT coordinator, urban school, 2002).

For a number of ICT coordinators, the time the role was taking meant they were having to consider giving it up.

Time is a real issue, is a huge issue with me. To be fair, something's got to give, and its either my role as a DP [deputy principal] or my role as a teacher, or it's the de facto role of ICT coordinator (ICT coordinator, rural school, 2002).

I'm almost at the stage where I'm probably thinking of giving it away because it's frustrating, it's too much hassle and I'm not getting the time to do it so, I'm getting hassled by people about it and, cripes I can't be bothered with this. I think you need half a load really if you want to do a decent job, if you want to be able to get around into teachers' rooms and help them out (ICT coordinator, provincial school, 2002).

ICT coordinators reported being put in the difficult position of having their responsibilities to their classes pitted against their responsibilities as ICT coordinator for the school as a whole. One ICT coordinator spent almost an entire week away from his classes after their school got a computer virus, while others faced similar issues: 'My other big issue with that is the number of times that I still get called out of class because little Johnny's got a problem, or whatever it might be. And I have to leave my class – to go and fix that problem. So my classes are getting a raw deal out of it' (ICT coordinator, provincial school, 2002).

Some coordinators had never had their role officially recognised. As one commented, 'I've never actually been officially appointed ICT – its only been done by default because ... I've had an interest in computers and then I was seen as being the main architect with the ICT plan and then most of the physical cabling has been sort of something which I've dragged out cables and things like that, and ... I had that understanding of how it all fits together and when things fall over I become the technician (ICT coordinator, rural school, 2002).

The importance of acknowledgement and recognition of those with roles in ICT was also mentioned by another coordinator: 'There isn't a real forum for discussion because we don't have a curriculum committee at this school and last year we did have sort of ICT curriculum committee meetings but they were not particularly productive and it would probably be better if it was a more formal, more recognised thing that was part of the school' (ICT coordinator, urban school, 2002).

Professional Development

Another issue which faced the coordinators in these schools was their own knowledge, in terms of both ICT skills and its integration into teaching and learning. They had become ICT coordinators through a variety of paths, and had varying levels of knowledge and experience with regards to ICT: 'So I'm not employed here specifically as a computing teacher. I just have computing as a background prior to teaching so that's why I ended up in the role I'm in ... my hand maybe went up higher than everybody else's when, when questions are asked and then I was tapped on the shoulder (ICT coordinator, urban school, 2002).

Several coordinators commented that they needed PD, either in specific skills such as networking or, more commonly, in terms of how ICT could be used effectively in all the subjects taught in their school. In 2002, the majority of ICT coordinators indicated that they would very much like the opportunity to visit other schools to see exemplary practice in ICT use (80 per cent) and to attend conferences on teaching and learning with ICT (79 per cent). In contrast, only half wanted to undertake high level network training courses, and 35 per cent wanted to attend conferences focusing on technical aspects. What they needed were 'IDEAS ... [and] examples of good practice' (provincial school). 'I personally believe there's a big PD issue here. That we can make people happy and comfortable using word processor and email and the Internet and all those things. But what we've actually got to ... not give them, but to show them, is how that can be incorporated into your day-to-day teaching. And that I still think is the hard thing' (ICT coordinator, provincial school, 2002).

Conclusion

It would appear, then, that ICT coordinators in general, and in these schools in particular, were in the position to be able to lead changes involving ICT to enhance teaching and learning. Generally, they were already doing this to at least some degree; however, a lack of recognition and support for their role as leaders impacted on their ability to do so. Strudler's (1994) study on the role of technology coordinators in US schools led him to recommend that schools should consider staffing on-site technology coordinators if they wish to successfully integrate technology into the school curriculum. He maintains that without the time allocation for the coordinator to provide the leadership to establish a shared vision and develop a school plan, integrating ICT into the school curriculum is unlikely to occur. Having an ICT coordinator fulfil this role, although not sufficient to ensure the effective integration of ICT within schools, is a necessary factor (Becta, 2002), and one that must be addressed. It would appear, then, that the leadership role of ICT coordinators needs recognition at both the school and national level. At the school level, the principal and senior management need to ensure that the role and responsibilities of the ICT coordinator are clearly understood by all members of staff, while funding is needed at the national level to ensure ICT coordinators have the time to fill this role, and to undertake the professional development required to do so. In addition, there must be recognition that many ICT coordinators are spending significant portions of their time on technical and/or administrative support, even though this is not where they are best utilised. It would seem that the routine technical and administrative aspects of ICT use in schools should be dealt with by people with experience and expertise in these areas, leaving the coordinators who are experienced teachers to focus on their area of expertise, the use of ICT in teaching and learning.

References

Ang, C. (1998). *Evaluation report: The FY98 professional development days (PDD program)*. Available at: <http://www.palmbeach.k12.fl.us/9045/pdd.htm>

Becker, H.J. (1999). *Internet use by teachers: Conditions of professional and teacher directed student use. Report #1.* Center for Research on Information Technology and Organisations, University of California, Irvine and The University of Minnesota (February). Available at: <http://www. crito.uci. edu/TLC/findings/internet-use/>

Becta (2000). *A preliminary report for the DfEE on the relationship between ICT and primary school standards: An analysis of Ofsted inspection data for 1998–99.* Available at: <http://www.becta.org. uk/research/resources/ictresources.html>

Becta (2002). *Primary schools – ICT and Standards: A report to the DfES on Becta's analysis of national data from OFSTED and QCA.* Available at: <http://www.becta.org.uk/research/reports/ictresources. html>

Bruder, I. (1990). The Third Computer Coordinator Survey. *Electronic Learning, 9*(7), 24–29.

CEO Forum (1999). *Professional development: A link to better learning.* The CEO Forum school technology and readiness report. Available at: <http://ceoforum.org/reports.cfm?CID=2&RID=2>

Cook, C.J. (1997). *Critical issue: Finding time for professional development.* Available at: <http://www. ncrel.org/sdrs/areas/issues/educatrs/profdevl/pd300.htm>

Education Review Office (2001). *The implementation of information and communications technology (ICT) in New Zealand schools 2001.* Available at: <http://www.ero.govt.nz/Publications/pubs2001/ ICT%202001.htm#Title>

Falloon, G. (n.d.). Developing Exemplary Practice: Why are some teachers better at IT than others? *Unpublished manuscript.*

Fullan, M. (1991). *The new meaning of educational change.* London: Cassell.

Fullan, M.G. (1992). *Successful school improvement.* Buckingham: Open University Press.

Fullan, M. (1993). *Change forces: Probing the depths of educational reform.* London: Falmer.

Fullan, M.G. (2001). The NEW meaning of educational change (3rd edn). London: Routledge Falmer.

Glennan, T.K. & Melmad, A. (1996). *Fostering the use of educational technology: Elements of a national strategy.* Available at: <http://www.rand.org/publications/MR/MR682/contents.html>

Hargreaves, A. (1994). *Changing teachers, changing times: Teacher's work and culture in the postmodern age.* London: Cassell.

Hopkins, D., Ainscow, M. & West, M. (1994). *School improvement in an era of change.* London: Cassell.

Lai, K.W. (1999). Teaching, learning, and professional development: The teacher matters most. In K.W. Lai (ed.). *Net-working: Teaching, learning, & professional development with the Internet* (pp.7–24). Dunedin: University of Otago Press.

Lai, K.W. (2001). Role of the Teacher. In H. Adelsberger, B. Collis and J. Pawlowski (eds), *Handbook on information technologies for education and training.* Berlin: Springer-Verlag, (343–54).

Lai, K.W., Elliot, A.G. & Trewern, A. (1999). Ethical use of computers in New Zealand Schools: A preliminary study. In *Proceedings of ICCE '99: 7th International Conference on Computers in Education*, 648–51.

Lai, K. & Pratt, K. (2002). *Primary Technology Project: Evaluation of the use of technology in Otago primary schools.* Dunedin: Community Trust of Otago.

Lai, K.W. & Pratt, K. (2004). Information and communication technology (ICT) in secondary schools: The role of the computer coordinator. *British Journal of Educational Technology, 35*(4), 461–75.

Lai, K.W., Pratt, K. & Trewern, A. (2001). *Learning with technology: Evaluation of the Otago Secondary Schools Technology Project.* Dunedin: Community Trust of Otago.

Lai, K.W., Pratt, K. & Trewern, A. (2002). *e-Learning initiative: Current state of ICT in Otago secondary schools.* Dunedin: Community Trust of Otago.

Lai, K.W., Trewern, A. & Pratt, K. (2002). Computer coordinator as change agents: Some New Zealand observations. *Journal of Technology and Teacher Education, 10*(4), 539–51.

Lawton, D. (1994). Defining quality. In P. Ribbins and E. Burridge (eds), *Improving education: Promoting quality in schools*. London: Cassell, (1–7)

Lucock, S. & Underwood, G. (2001). *The role of the ICT coordinator*. Available at: <http://www.pfp-publishing.com/primary/ict-cont.htm>

Marcovitz, D. (2000). The roles of computer coordinators in supporting technology in schools. *Journal of Technology and Teacher Education*, 8(3), 259–73.

Means, B. & Olson, K. (1995). *Technology's role in education reform: Findings from a national study of innovating schools*. Available at: <http://www.ed.gov/PDFDocs/techrole.pdf>

Moallem, M., Mory, E. & Rizzo, S. (1996). *Technology resource teachers: Is this a new role for instructional technologist?* ERIC Reproduction Service No. ED 397 823.

Moursund, D. (1992). *The technology coordinator*. Eugene, OR: International Society for Technology in Education.

Murray, D. & Campbell, N. (2000). Barriers to implementing ICT in some New Zealand schools. *Computers in New Zealand Schools*, 12(1) 3–6.

National Education Association [NEA] (1999–2000). *It's about time*. From the 1999-2000 New Member CD. Available at: <http://www.nea.org/bt/3-school/time.pdf>

November, A.C. (1990). Guest editorial: The emerging role of the computer coordinator. *Electronic Learning*, 9(7), 8–9.

Reilly, R. (1999). The technology coordinator: Curriculum leader or electronic janitor? *MultiMedia Schools*, (May/June), 38–41.

Ringstaff, C., Yocam, K. & Marsh, J. (1995). *Integrating technology into classroom instruction: An assessment of the impact of the ACOT Teacher Development Center project*. ACOT Report #22. Available at: <http://www.apple.com/education/k12/leadership/acot/library.html>

Ronnkvist, A., Dexter, S.L. and Anderson, R.E. (2000). *Technology support: Its depth, breadth and impact in America's schools* (Teaching, Learning, and Computing: 1998 National Survey. Report #5.). Available at: <http://www.crito.uci.edu/tlc/findings/technology-support/startpage.htm>

Schiff, T.W. & Solmon, L.C. (1999). *California Digital High School Process Evaluation Year One Report*. Milken Family Foundation for the California Department of Education. Available at: <http://www.mff.org/publications/publications.taf>

Sheingold, D. and Hadley, M. (1990). *Accomplished teachers: Integrating computers into classroom practice*. New York: Bank Street College of Education.

Smerdon, B., Cronen, S., Lanahan, L., *et al*. (2000). *Teachers' tools for the 21st century* (No. NCES 2000-102). Washington, DC: National Center for Education Statistics, US Department of Education.

Somekh, B., Barnes, S., Triggs, P. et al. (2001). *NGfL Pathfinders: Preliminary report on the roll-out of the NGfL programme in ten pathfinder LEAs*. Available at: <http://www.becta.org.uk/research/reports/ict-re.html>

Strudler, N. (1994). *The role of school-based technology coordinators as change agents in elementary school programs: A follow-up study*. ERIC Reproduction Service No. ED 381 139.

Strudler, N. & Gall, M. (1988). *Successful change agent strategies for overcoming impediments to microcomputers implementation in the classroom*. ERIC Reproduction Service No. ED 298 938.

Strudler, N., Falba, C. & Hearrington, D. (2001). *The evolving role of school-based technology coordinators in elementary programs*. Paper presented at the National Educational Computing Conference: Building on the Future, Chicago, IL, July 25–27. Available at: <http://confreg.uoregon.edu/NECC2001/program/research_pdf/Strudler.pdf>

Improving Educational Opportunities through ICT Partnerships

6

Linda Selby, Ken Ryba and Garry Falloon

A significant aspect of ICT has been the opportunities that it has provided for the formation of partnerships between education, business and community. This is evident, for example, through recent work that has been undertaken in Auckland with the establishment of the North Shore Schools Net and the South Auckland Schools Net projects (Ryba, *et al.*, 2002; Ryba, *et al.*, 2003). The establishment of partnerships with families, businesses and the wider community is one of the goals that has been set by the Ministry of Education in the *Digital Horizons: Learning through ICT* policy document (Ministry of Education, 2003). This goal stresses the sharing of knowledge about ICT and the extension of opportunities for learning through ICT. Strategies for schools, government and other stakeholders to achieve this goal are to:

1. strengthen links between home, schools and communities through ICT initiatives and enhancing public understanding of how children learn and how ICT can enrich learning environments;
2. align Ministry programmes with regional and community ICT initiatives that extend opportunities for learning through ICT and building partnerships with agencies;
3. work with businesses to understand changing workplace needs and patterns, and to extend partnerships to enhance ICT use in schools; and
4. encourage government agencies and business to donate surplus computers to schools through recycling schemes (Ministry of Education, 2003, p. 19).

A feature of the above strategy is that it provides reciprocal benefits for the various partners and places schools on a level playing field with business and community agencies. The growing interest in partnerships with ICT has been fuelled by recognition that education is not simply a government responsibility, but that the future of our nation will depend upon our ability to work in partnership with families, businesses and the wider community. The quality of communities is defined by the strength of its schools. While it is obvious that the most important stakeholders are students and parents, local employers and other businesses also have a vested interest in the success of education within their community. In the past, businesses have made various philanthropic contributions to school programmes and activities through donations of money, naming rights, and voluntary support. Present approaches, however, tend to recognise the potential benefits that can accrue through the formation of business, community and education partnerships that are based on a shared understanding of values and culture to support mutual needs (Council for Corporate and School Partnerships).

School–business relationships in New Zealand have been a focus of attention by the Educational Review Office (ERO) for several years now (ERO, 1996). The involvement of business in the development and delivery of educational programmes has its roots in transition and work-experience programmes that historically targeted non-academic students who were likely to enter the workforce directly after leaving school. It has been common practice in secondary schools to invite people from business and industry to speak to students about career opportunities. ERO notes, however, that there have been tensions between the school and business sectors. Business leaders have often been critical of aspects of education, including curriculum content, the qualification system and standards of student achievement. In turn, educational leaders have often resented the imposition of business-sector management models, systems and principles into school decision-making, particularly since the advent of Tomorrow's Schools reforms.

With the above factors in mind, the purpose of this chapter is to: (1) explain the guiding principles and underlying foundations of partnerships; (2) illustrate effective partnerships through case study examples; and (3) discuss the implications and issues of partnerships with ICT as this affects the formation of relationships between businesses, schools and communities.

Guiding Principles and Policies for School–Business Partnerships

The Council for Corporate and School Partnerships has identified some guiding principles for business and school partnerships. These state that partnerships must:

- be based on shared values and philosophies;
- be defined by mutually beneficial goals and objectives;
- involve activities integrated into the school and business cultures;
- be driven by a clear management process and structure;
- be defined by specific measurable outcomes;
- be supported at the highest level within the business and school so that there is concurrency at all levels;
- be founded on detailed internal and external communication plans that clearly illustrate expectations of all parties;
- be developed with clear definitions of success for all partners.

A *Framework for Developing School–Business Community Relationships* was announced by the Ministry of Education, Victoria, Australia in 1998 (Victoria State Department of Education, 1998). This was in recognition of the importance of going 'beyond the gate' to enhance educational opportunities for students. A key feature of this policy is a clear statement indicating that while such partnership ventures are desirable, they must not compromise the values of the school or its leadership and management independence. Under the policy:

- all school–community/business partnerships must directly contribute to the enhancement of educational opportunities for students or have some educational purpose;

- a copy of all documents relating to the relationship must be made available to the Department of Education for audit purposes;
- such partnerships should not restrict the choice of parents and students in the purchasing of goods and services;
- school–community/business partnerships should be limited to persons who can demonstrate an involvement that will contribute to the educational purposes of the school and the values, goals and policies of the department and the local school community;
- arrangements with schools of a direct nature must not be entered into with companies involved with tobacco, alcohol or gambling, offensive materials or themes
- arrangements must comply with the national competition code;
- school council approval is required for all arrangements between schools and outside bodies.

Examples of Effective ICT Partnerships

Digital Opportunities initiatives

Digital Opportunities is a trial initiative developed in partnership between business, government and education. This initiative is coordinated by Graeme Plummer, the project manager of Digital Opportunities at the Ministry of Education, and Wiremu Grace, te reo Maori content coordinator. Established to explore new, innovative ways to encourage teachers and students to communicate and participate online, it is providing the means for schools to more effectively share their unique experiences, knowledge and resources within online e-learning environments.

Digital Opportunities is made up of four projects:

GenXP: In this project, schools in Gisborne and West Auckland are providing new opportunities for their students to gain qualifications in Microsoft applications.

FarNet: Teachers and students of ten Northland schools are linked through this bilingual e-learning community where they are sharing an increasing range of resources and collaborating in online activities.

WickED: Developed for the students, and facilitators, of four study support centres in Christchurch and Invercargill, WickED is a quality assured, lively learning environment, hosted by virtual characters Ed and Wiki, and full of student-friendly activities and interactives.

Notebook Valley: Through this project, senior secondary school students of three Lower Hutt schools have been provided with notebook computers and Internet access to assist them in their study of science, maths and technology.

The Internet is a growing force in learning and the Digital Opportunities initiative provides the means to explore how it can be harnessed and integrated in the everyday teaching and learning lives of New Zealand schools. The projects include a diverse range of activities and materials, for example:

- teacher-produced curriculum materials and resources;
- an online collaborative exploration of the geology of the regions;
- student magazines written and published by the participating students;
- the stories and memories of the local kaumatua, published in te reo Maori;
- the interactive games and activities of WickED, published in English and te reo Maori to support literacy and numeracy outcomes for young people, and to support the teaching of science, maths and technology.

The projects continue to expand and grow as the participating students, teachers and communities find new ways to collaborate and share online.

The Ministry of Education has actively pursued a number of initiatives to ensure the optimal development of ICT in support of teaching and learning. This has included:

- ICTPD clusters;
- ICT advisory services;
- improved Internet access;
- principals' laptop, web portal and online network;
- laptops for secondary school teachers;
- laptops for primary school teachers;
- ICT Helpdesk for schools;
- Te Kete Ipurangi – The Online Learning Centre.

Descriptions of these projects are available online <www.minedu.govt.nz>.

The common purpose of these initiatives has been to increase knowledge, skills, and resources among schools and clusters of schools so that they can make effective use of ICT. Beyond such school-based initiatives, however, there is a need for new ways of working across sectors to increase the capacity of ICT use in schools. With this purpose in mind, North Shore Schools Net (NSS-NET) came into being as an alliance between a university, a college of education, schools and business interests.

North Shore Schools Net (NSS-NET)

Mindful of the principles and policies of school–business/community partnerships NSS-NET sought to develop a unique approach that would be more responsive to the needs identified in each school and to help the schools establish ICT projects through bottom-up rather than top-down initiatives. A cross-sector alliance between schools, tertiary institutions, business and community was established. The NSS-NET undertook to set up a knowledge network among schools working on related projects in several primary and secondary schools. The vision of NSS-NET was to 'increase the capacity of participating schools to effectively apply ICT in order to create better conditions for teaching and learning through shared professional development, technical support, and the provision of equipment and resources for applied research and projects'. This was accomplished through an alliance of four pilot schools, the Tindall Foundation, Massey University, Team Solutions and the Centre for Professional Studies, Auckland College of Education (ACE).

The participating schools were: Kauri Park School, St Joseph's School, Rosmini College and Carmel College, all on Auckland's North Shore. Schools were responsible for identifying their project and for carrying it through to completion with the support of the alliance. Following is a brief description of the projects that were undertaken by schools (for a detailed description of each project, see the November 2002 issue of the *Computers in New Zealand Schools* journal).

Kauri Park: The purpose of this project was to create a library information centre. This involved the redevelopment of the physical space of the library and appropriate professional development of staff to increase their skills in using information to enhance teaching and learning across the curriculum areas.

St Joseph's School: The purpose of this project was to set up a system to carry news broadcasts accessible to classrooms, the school community and beyond. The broadcasts are linked to the school website so that this information can be available to other schools and the wider community.

Carmel College: The aims of this project were to increase environmental awareness by including the school and community in gathering information on the health of waterways in the local environment in and around Lake Pupuke. This was an integrated studies project in which students across a range of subject areas worked together to collect and analyse the data.

Rosmini College: The Fitnet Project involved all students in the measurement, tabulation, analysis and reporting of their own personal and group fitness scores. The aim was to provide students with a lifelong skill to measure and improve their personal fitness and health. Students had opportunities to compare their skills with similar-aged students from around the world.

In addition to the above projects, Carmel College and Rosmini College trialled performance software called Gamebreaker™. This software is a digital video-analysis programme that can be used as a performance analysis tool for any field of sports and professional performance. The suppliers of the software are supporting this trial in order to explore the educational potential of this application as a teaching tool rather than a coaching tool.

What Was Different about North Shore Schools Net?

Cross-sector alliance between schools, business and tertiary institutions: A distinguishing feature of NSS-NET has been the creation of a professional community of practice in which planning, implementation and evaluation of projects are integrated as a shared problem-solving activity. Action research was selected as a framework for developing and evaluating the projects on the basis that this approach could promote a working partnership of teachers, university, and project staff.

Mutual benefits: The belief underlying this approach was that partners would derive mutual benefits from working collaboratively with one another.

Figure 1: Benefits of NSS-Net.

Benefits for schools	Benefits for other partners
• Funding of hardware and software for development projects in each school	• Massey University and ACE benefited from increased enrolments resulting from delivery of courses and workshops to schools
• Subsidies for course fees and professional development costs	• The Tindall Foundation, Massey and ACE received recognition for their roles in creating better conditions for learning with ICT in schools
• Networking with other funding agents and Ministry of Education initiatives	
• ICT facilitation, professional development and training delivered to the school	• Massey University, ACE and schools worked together to link research and theory to innovative ICT projects
• Technical advice and support on hardware and software	
• On-line support and resource sharing	• All partners benefited through an increased public profile of their ability to work together in order to increase the capacity of schools to make effective use of ICT in support of teaching and learning
• Participation in Massey and ACE training, workshops and conferences	
• Participation in Massey and ACE courses and qualifications	
• Whole-school collaborative approaches to ICT development	

Building a community of professional practice

Within the NSS-NET Project, a 'community of professional practice' was created in which participants developed an identity as members of the community and worked collectively to achieve the aims of the project. The formation of this community can be thought of as a learning journey in which the teachers gain knowledge and skills in the process of working together as a team with academics, researchers, and technical advisers (Lave & Wenger, 1991). This was evident in the following activities:

- preparing a CD-Rom video to showcase the school projects at an official launch;
- working together in an action research team to document the development and outcomes of their projects;
- participating in lead-teacher meetings to review progress and plan each stage of the project;
- preparing a collaborative progress report for the financial sponsor of the project;
- hosting a final presentation of the project results and providing information to other schools;
- writing articles on the action research projects for publication.

Four main dimensions contribute to the formation of a vibrant and sustained community of professional practice (Hung and Chen, 2001). These dimensions are briefly described below:

Situatedness: learning was embedded within the NSS-NETproject through action research and collaborative work aimed at developing a more 'global picture'. The action-research process encouraged teachers to reflect on their actions through discussion of issues and problems with fellow community members.

Commonality: participants worked together in ways that made sense to themselves. They shared interests and problems that required joint effort and, in the process, developed a similar bonding or identity with one another. Commonality was not only socially mediated, but it included a common framework for planning, implementing and researching the projects (Lave and Wenger, 1991).

Interdependency: participants interacted with one another based on varying needs and levels of knowledge and skills. They made use of one another's abilities to increase their own understanding and professional skills. Participants developed in areas where they were most interested and capable with a responsibility for sharing their understanding with the other participants within the community.

Infrastructure: NSS-NET made use of a common framework for carrying out and researching the projects. This common framework enabled teachers within the community to share their knowledge and understanding. Such a framework enabled participants to be structurally dependent on one.

Project framework

Collaborative development: The members of the alliance worked together to facilitate ICT developments in each of the pilot schools. The NSS-NET manager/ facilitator worked with each school to assist them in the development of projects aimed at meeting specific teaching and learning objectives.

Project manager/facilitator: The manager/facilitator worked collaboratively with school principals and selected staff to develop proposals that were submitted to the NSS-NET management committee for funding consideration.

Memorandum of understanding: Selected schools became strategic-alliance partners and a memorandum of understanding was prepared outlining the aims and responsibilities of the partner members. Principals were asked to represent their schools in order to establish a school-initiated approach.

Development fund: One project was funded in each of the participating schools. A procedure was developed to match the distribution of funding to the needs of schools. There were set criteria and a protocol for making a proposal for funding. Decisions on funding were made by the NSS-NET Executive Committee.

School-based initiatives: The aim was to encourage school-based initiatives by funding projects that enabled schools to undertake specific development projects that were justified in terms of teaching and learning needs in each school. Projects took many different forms, depending on the needs of each school.

Funding and support: Support for projects included: (1) funding of hardware and software; (2) ICT facilitation; (3) technical support; (4) professional development; and (5) course tuition fees and other project related expenses. The proposal process was flexible so that schools could gain assistance in ways that matched their needs. Schools were required to demonstrate that they could match the funding request on a one-to-one basis with equivalent funding from their own budget. Each school provided a detailed financial breakdown of expenses in order to show what their contributions to the project were in direct proportion to the funding request.

Support for Schools
Manager/facilitator: The manager/facilitator worked alongside the key teachers from the four schools to identify and interpret a vision for teaching and learning that moved away from the rhetoric of ICT to practical application. Many meetings were held to discuss how to implement the vision of 'increasing the capacity of participating school to effectively apply ICT to create better conditions for teaching and learning'. Each school had its own specific needs. It became clear that a framework had to be developed that allowed schools to choose their own directions while still adhering to the vision.

At the commencement of the project, the manager/facilitator worked in the schools with key teachers. She helped schools submit a proposal to the executive committee for approval. The selected projects directly benefited the students and school. She made weekly contact with schools, had practical input at a staff development and pupil training level, gave assistance in sourcing equipment required by each school and advice on ICT integration into the curriculum. Liasing with the schools, the executive committee, Massey University, ACE, Team Solutions, the Tindall Foundation and commercial firms meant communication was established between all participants.

Specialised software: Teachers and students were taught how to use the following specialised software: I-Movie multimedia, Powerpoint presentation, Dreamweaver MX/web development, Data logging, Pure Voice/digital audio and Excel spreadsheet.

Web development: The School of Mathematics and Information Sciences at Massey University, Albany Campus, provided training, consultation and ongoing support for teachers and students who created websites for their projects. A total of eight senior university students worked on-site in the schools with the aim of assisting school staff and students to gain the knowledge and skills needed in order to independently manage and maintain their own websites. Dreamweaver MX/web development software was purchased for all schools for this purpose. The web development aspect of the project was supervised by Associate Professor Scott Overmyer, School of Mathematics and Information Sciences.

NavCon Conference 2K2: Schools were invited by the NSS-NET executive committee to send key teachers to the NavCon Conference in Christchurch. The conference provided an opportunity to find out about new initiatives and best practice in ICT in Australia, New Zealand and beyond.

Project management
The NSS-NET executive committee managed the overall project. The role of executive committee was to: establish the criteria for approval of projects; assess

applications and determine suitability in relation to criteria; be financially accountable to the funding agents; provide advocacy for schools, promote, support and motivate to ensure programme success and longevity; handle media relations; and manage accounting and reporting procedures.

The unique feature of NSS-NET was the cross-sector cooperation that was achieved through forming an alliance of business, schools and tertiary-education institutions. Cross-sector partnerships are important not only for resourcing the development of ICT in schools but for ensuring the relevance of education to the growth of knowledge in society. NSS-NET illustrated how such partnerships can work in practice to the benefit of all participants.

South Auckland Schools Net (SAS-NET)

The South Auckland Schools New Educational Technologies project (SAS-NET) came into existence as an extension of our work on the NSS-NET. Following the same framework for supporting school-based initiatives, the vision of SAS-NET was to: 'increase the capacity of participating schools to effectively apply ICT in order to create better conditions for teaching and learning through shared professional development, technical support, and the provision of equipment and resources for applied research and demonstration projects'.

This was accomplished in South Auckland through an alliance of a cluster of seven schools, the Tindall Foundation, Massey University College of Education, Team Solutions and ACE's Centre for Professional Studies.

Schools were responsible for carrying out their projects with support of the alliance. A lot of groundwork for ICT integration into classroom practice had been undertaken by this cluster of schools in preparation for an application to the Ministry of Education for ICTPD funding. The SAS-NET executive committee considered that participation in the project by this particular group of schools could position them well for future funding opportunities.

Following is a brief description of the projects that were undertaken by schools. (see *Computers in New Zealand Schools*, November 2003, for a detailed description of each project). All participating schools placed a strong emphasis upon teacher professional development and linked this goal with their school projects.

Park Estate Primary: This project aimed to utilise film-making as a means of enhancing students' visual literacy skills through the use of digital video and digital still photography. The school also wanted to increase the extent to which it was using ICT across the curriculum and as a means for communicating student achievement to parents and the school community.

Papakura South School: This Maori medium school considered it important to be able to access quality education through interactive ICT applications that could be related to their tikanga context. The project initially involved the use of webquests and then advanced to the use of digital animation using Claymation™ to engage the students in the retelling of Maori stories.

Edmund Hillary Primary: As a Maori medium school with both bilingual and immersion programmes, this project focused on visual literacy through digitally recording traditional stories in Te Reo Maori. The aim was to assist the development of language and visual literacy through re-telling stories in a video film format.

Kelvin Road Primary: The aim of this project was to establish a digital daily news broadcasting programme with a focus on video editing. Through the active involvement of teachers and students, video capture and editing was set up so that it could be used in any classroom or area of the school.

Mansell Senior School: This project was about enabling the integration of ICT into teaching and learning by developing digital literacy through the use of video and still photography. The highlight of this project was the production of an advertisement that made the finals of television's 'Fair Go' Ad Awards.

De La Salle College: The aim of this project was to establish an intranet that could be used across all subject areas, with the possibility of extending it beyond the boundaries of the school to enable student and community access from home.

Mcauley High School: This project focused on the use of ICT as a context for developing a healthy schools programme. Students and teachers developed the programme together with a focus on improving the health, fitness, and diet of students in the school. This is a longitudinal study that will involve the ongoing monitoring of health factors in order to assess the effectiveness of the approach.

Unified learning approach to ICT adopted within SAS-NET

ICT has provided a rich ground for the development and refinement of theories and metaphors about the teaching and learning processes that guide our work as teachers, students and researchers. Two such metaphors that have gained popularity within education in recent years are the 'acquisition' metaphor and the 'participation' metaphor (Sfard, 2003). In basic terms, the acquisition metaphor is concerned with how knowledge is constructed by the student, whereas the participation metaphor is concerned with the social situation and context within which the knowledge is constructed. Emphasis within the acquisition metaphor is placed on the development of individual enrichment and learning strategies. In contrast, the participation metaphor stresses the formation of learning communities and the mechanisms for generating knowledge through community-building processes.

Various writers have put each of these two metaphors forward as a useful way for understanding how to use ICT as a context for learning. However, rather than becoming devoted to one metaphor or the other, our view is that it is best to adopt a unified learning approach that incorporates both an emphasis on the individual learning process as well as approaches for building learning communities. With this unified approach in mind, we were able to incorporate both metaphors into the SAS-NET project.

The advent of ICT in education has reinforced the adoption of theories and beliefs about 'knowledge acquisition' and the notion of children as developers of their own intellectual structures (see Papert, 1980). There are many terms that have been applied to denote the action of one's own acquisition of knowledge, including: reception, construction, attainment, development, accumulation (Sfard, 1998). Working within this acquisition model, the role of the teacher is to facilitate, guide, mediate and coordinate the learning activities. Once acquired, the knowledge can be applied by the individual, transferred to other contexts and shared with other learners. According to Sfard (1998), it is the metaphor of acquisition that underlies much of our thinking about constructivism and the notion of ownership over personal knowledge and some kind of self sustained entity.

Within the SAS-NET project, many of the teachers undertook PD in information literacy. This was provided through participation in the *Infolink: information literacy skills* course offered by the Auckland College of Education. Virtually all of the projects undertaken placed emphasis on individual learning processes and tasks for which computers are best suited (e.g. accessing and organising information, displaying and communicating information visually). The creative use of ICT for such purposes as animation and digital storytelling has the potential to significantly impact on learning and personal development (Means *et al.*, 1993).

Table 1 provides a summary of the two main metaphors for learning. The purpose of this table is to show how the two metaphors complement one another on each of the dimensions shown in the centre column. In other words, the individual learning cannot be separated from the context in which it happens. Increasingly, attention is being given to the use of ICT as a context for building a community of learners in the ways evident within the projects undertaken by the individual schools.

Table 1: Two metaphors for learning (adapted from Sfard, 1998, p. 7).

Acquisition metaphor	Dimension	Participation metaphor
Individual learning processes	*Goal of Learning*	Community building
Acquisition of specific skills and knowledge	*Learning*	Becoming a participant
Recipient, consumer and (re)constructor	*Student*	Peripheral participation, apprentice
Provider, facilitator, mediator	*Teacher*	Expert participant, preserver of community practices and discourse
Property, possession, commodity	*Knowledge, concept*	Aspect of practice and application of discourse within community activities
Having, possessing knowledge and information	*Knowing*	Belonging, identifying, participating, communicating

The unified approach has several underlying principles that are important for guiding the development of school-based ICT projects:

- Focus on the learning process through information literacy and metacognitive skill development within the context of a community of learners.
- Check to ensure that students acquire knowledge and skills in a timely way at the point where they need these skills to advance their learning through active participation.
- Provide students with opportunities to be both consumers and constructors of knowledge. The role of the teacher is to scaffold the learning process with strategic guidance and intervention rather than relying upon incidental learning.
- The role of the teacher is to provide, facilitate and mediate as an expert participant and as a guide working alongside students.
- Develop a 'language to think with' and problem-solving frameworks (e.g. storyboards) to scaffold the development of knowledge and concept acquisition.
- Provide opportunities for students to test their ideas and reflect on their thinking in an accepting environment where they feel a sense of belonging.

The above principles were put into practice in several ways through: training and development of an action research process; teacher professional development in information literacy; projects and applications that enabled active participation of students; advice and guidance from the project facilitator on how to scaffold student learning; and production of articles by the collective of teachers working to share information with others on the outcomes of their school-based projects.

The aim of NSS-NET and SAS-NET was to demonstrate what can be accomplished through cross-sector cooperation and the formation of an alliance between business, schools and tertiary education institutions. Cross-sector partnerships are important not only for resourcing the development of ICT in schools, but for ensuring the relevance of education to the growth of knowledge in society.

Initially, it was our intention that the Ministry of Education might look to advance the schools-net model. However, indications are that the SAS-NET project approach falls outside the current Ministry ICT in Schools initiatives. After completion of a two-year pilot project, it was the decision of the Tindall Foundation to conclude their support for school-based ICT initiatives. As an alternative, the Tindall Foundation plans to consider funding new and innovative projects that would not normally be provided by the Ministry for particular school projects. It is our hope that SAS-NET and other similar projects serve as an example of what can be accomplished through business and tertiary education partnerships with schools and how such partnerships can work to the benefit of all participants.

(Note: This school cluster has just been successful in its application for the Ministry of Education ICTPD funding for 2004–2006.)

Aligning ICT Partnerships with Educational Goals

Relationships between business and educational organisations need to be well managed and clearly linked to specific educational goals. This is important to ensure that the capacity building within the school as a result of partnerships continues long after the project finishes. The overarching guiding principle that should inform the development of partnerships is the educational well-being of the student (ERO, 1996). Working with this aim in mind, partnerships should be seen as initiatives and opportunities to create better conditions for student learning. All parties need to be involved at every stage in the project but particularly in the initial planning. It is also important that a 'bottom-up' rather than a 'top-down' approach is used because it is the teachers and students in the classroom who have to implement and sustain the project. Accordingly, it is essential that school staff are actively involved in decision making and that they have ownership of the project. Ultimately, it is the initiative of teachers at the chalk face that will bring about positive innovation and change through such partnerships.

Following are the main indicators of effective partnerships:

Clear statement of purpose in relation to learning goals: The partnership itself is founded on definite statements of purpose and expected learning outcomes defined with reference to the New Zealand curriculum documents. This includes justification of a particular project with reference to specific learning goals and learning areas of the curriculum. For example, in the Carmel School Project outlined above, the goals were clearly related to specific aspects of secondary science curriculum (logging data, analysing and presenting results of scientific experiments).

It is important to ensure that the concept of the school–business partnership is understood by all those involved in their development. School and business management must be clear about the rationale for such links. There needs to be a clear understanding of how school–business links fit with the broader vision and values of the school. This can be accomplished through the development of a statement of purpose and memorandum of understanding.

Management level support from school and business: School–business relationships are most likely to be successful and sustainable when they are supported at management level within both the school and the business. This is important for ensuring that the formation of the partnership is part of the overall operation of the school and not dependent for its success on a single teacher or a department working in isolation. Links between business and specific departments of the school may be vulnerable to changes in management and school priorities.

Manageable timeline and clear expectations of partners' responsibilities: The scale and time-frame of school–business partnerships should be realistic and achievable. This can be accomplished by having a reporting structure with milestone reports and dates for presentation of projects. The formation of a project committee can be an effective way of directly involving the partners and creating shared understandings and expectations. In the NSS-Net, the project committee consisted of representatives from the Tindall Foundation, Massey University, ACE, and schools. The committee ensured that project deadlines were met and that milestone reports and special events were organised on schedule.

It is important that the objectives of the relationship are clearly stated and the expectations of each party to the arrangement are discussed, agreed and recorded. When projects are well established, the success of each partner in meeting the objectives should be assessed, the overall initiative evaluated, and policy and procedures amended in the light of the findings (ERO, 1996).

The benefits for each partner should be clearly established along with shared understanding of the expected contributions from each party. Strategies for power sharing need to be considered (e.g. one:one funding or contributions in kind). Often schools underestimate their contribution to partnership initiatives in terms of the time and costs that they incur. Schools need to be selective in their choice of partners. This can be accomplished through drawing up profiles of possible partners and looking to opportunities for local partnerships before approaching large external corporations or trusts. Rethinking the importance of collaborative partnerships within business, education and the community means that schools no longer need to go 'cap in hand' as poor partners seeking donations from outsiders. There are reciprocal benefits to be achieved for all partners through effective dialogue, planning and collaboration.

Statement of educational outcomes and project outputs: A characteristic of successful projects is that they have a specific purpose that is defined by a starting point and by some definite outcomes achieved over a specific period of time. This was accomplished in the North Shore and South Auckland schools projects by hosting presentations at the beginning and end of the project and writing articles on each of the school projects. The specific outcomes were: (1) a launch at the outset of the project to which all relevant people from the business, school and community were invited; (2) a final multimedia presentation of achievements at the end of the project; and (3) a special issue of the journal *Computers in New Zealand Schools* presenting articles on each school project: project facilitators and teachers worked collaboratively to prepare the articles. A final report was also prepared detailing the specific outcomes of each project and how this related to the contributions of the various partners (students, teachers, business, tertiary-education organisations, and school administrators).

Barriers and Issues Concerning School–Business ICT Partnerships

A number of concerns about the formation of partnerships need to be considered. Many of these have been identified in a survey conducted by the Education Review Office on school–business links (ERO, 1996). Also, the Victoria State Department of Education framework for school–business/relationships that outlines the various problems and issues associated with the formation of partnerships (Victoria State Department of Education, 1998). The issues include:

Furthering of educational purposes: School/business/community partnerships must directly contribute to enhancement of educational opportunities for students. The purpose of projects needs to be carefully defined in terms of goals and objectives of the learning areas and essential skills contained in the *New Zealand Curriculum Framework* (Ministry of Education, 1993).

Power sharing: Partnerships need to be based on reciprocal benefits and should not adversely influence or advantage any one of the parties. Financial contributions should contribute directly to the good of the project as defined by the educational purpose. For example, it would be unacceptable for one party to seek pecuniary benefits or business favours from a business–educational partnership.

Choice: Partnerships should not restrict the choice of parents and students in purchasing goods and services. The participation of a specific business or agency should not imply endorsement for its products nor should it impact in any way on the right of parents and students to choose alternatives.

Location: Schools in isolated situations may have less opportunity to form partnerships within the local business community. Because there may be few local industries and a small population base, there may be advantage in clusters of small schools, industries and communities working together on a coordinated project that benefits all concerned.

Cost: Schools may not be well placed financially to contribute to the capital costs of setting up a partnership project, but they are well placed to contribute in other ways. These costs need to be calculated appropriately. The cost factor needs to be weighed up in terms of school budget priorities, staff capacity and the educational benefits that can be derived from participation.

Time: The establishment of school–business relationships can be time-consuming and the apparent benefits may not always justify the energy expended. Teachers and school administration need to commit to the development and implementation of a partnership project.

Attitude: Staff may be resistant to the ethics and values that are implicit in the involvement of business in education. There may be concerns about the motives of business and the influence that this could have on the development and delivery of educational programmes.

Competition for Collaboration: The current emphasis on collaboration within government policy and the Ministry of Education means that there is a great deal of competition among schools and other educational institutions to gain access to partnerships. Schools are competing among themselves and with other educational institutions for opportunities to participate in partnerships with business.

Conclusions

An essential aim of schooling is to prepare students to take an active and full role in the community and the workplace. ICT partnerships can provide a context for creating better learning conditions both within and beyond the walls of the school.

A feature of ICT partnerships is that they potentially give schools, businesses and communities an opportunity to work together in ways that contribute to the growth and education of our nation's children. The notion of education as a shared responsibility is evident in the various educational reforms that have been carried out over the past twenty years in Aotearoa New Zealand, stressing the interconnectivity of the community and the school.

School–business links are one means by which students, teachers and people in the business sector can enhance their understanding and appreciation of the role each has in the community. The examples presented in this chapter illustrate the reciprocal benefits that can result from effective ICT partnerships that are focused on enhancing educational opportunities.

It is important to plan partnerships carefully from the outset, in order to select appropriate partners and to develop shared understandings of the purpose, focus and intended educational outcomes of partnerships. The rethinking about partnerships and collaboration between schools, businesses and communities stresses the importance of schools viewing themselves as equal partners and active agents rather than as beneficiaries of external funding. This is essential to preserve the educational purposes of the school and the values, goals and policies of New Zealand education.

There are a number of issues and constraints that can impact on the formation of ICT partnerships. These need to be worked through in order to identify the most effective strategies for developing relationships that will add value to the educational process. Attitudes and beliefs of educators, business people and community members can play a significant part in determining whether or not partnerships are viable and sustainable. Current policies from the Ministry of Education encourage collaboration and the formation of relationships that contributes to the educational purposes of the school.

The increasingly strong emphasis on ICT partnerships and school–business relationships raises the question whether the Ministry of Education should be more proactive in advising, supporting, guiding, and funding the development and implementation of school–business relationships. Marsden (1994), a world authority in school–business relationships, has noted that policy-makers must ensure that the 'brokerage' process necessary to bring schools and businesses together is available and adequately resourced.

Finally, the formation of ICT partnerships should follow a bottom-up rather than a top-down process. School staff should be active agents of partnership ventures and must have a sense of ownership in the process. The selection of potential partners is a strategic process and the best results are likely to be obtained through the early involvement of all parties in the planning and implementation process. Shared understanding, clear expectations and definable outcomes that are related to specific educational goals are the key guiding principles of effective partnerships.

References

Council for Corporate and School Partnerships. Available at: <http:www.corpschoolpartners.org>.

Education Review Office [ERO], (1996). Available at: <http://www.ero.govt.nz/Publications/eers1996/96no2hl.htm#part6>

Hung, D.W.L. & Chen, D. (2001). Situated cognition, Vygotskian thought and learning from the communities of practice perspective: Implications for the design of web-based e-learning. *Education Media International, 38* (1), 4–12.

Lave, J. & Wenger, E. (1991). *Situated learning: Legitimate peripheral participation*. Cambridge: Cambridge University Press.

Marsden, C. (1994). *Building effective partnerships between schools and business: Response 2*. Paper presented at: Directions: Education and Training for 15–24 Year Olds (An international conference on policy imperatives and education opportunities for the education and training of 15–24 year olds). Sydney, 28–30 September.

Means, D., Blando, J., Olson, K. & Middleton, T. (1993). *Using technology to support educational reform*. Reports 20402–9328. Washington, DC: Office of Educational Research and Improvement.

Ministry of Education (1993). *New Zealand curriculum framework*. Wellington: Learning Media.

Ministry of Education (2002). *Overview of ICT programmes for schools*. Available at: <http://www.minedu.govt.nz>

Papert, S. (1980). *Mindstorms: Children, computers and powerful ideas*. Brighton: Harvester.

Sfard, A. (1998). On two metaphors for learning and the dangers of choosing just one. *Educational Researcher, 27*(2), 4–13.

Victoria State Department of Education (1998). *Framework for developing school-business/community relationships*. Available at: <http://www.robertclark.net/news/1018gude.htm>

Interaction and Control in Online Educational Communities

7

Bill Anderson

At the heart of a community of learners lies the communication that takes place between its members. That communication serves two main functions. In the broad terms that Gee (1999) uses discussing language, those functions are 'to scaffold the performance of social activities (whether play or work or both) and to scaffold human affiliation within cultures and social groups and institutions' (p. 1). Gee reminds us that these functions are connected; that ultimately 'language-in-use is everywhere and always "political" ' (p. 1); and that the political nature of language-in-use implicates it in the creation and distribution of social goods such as power, status and worth. In more specific terms relevant to online learning communities, the functions can be described for the student community members as scaffolding the attainment of the goals of their study, and scaffolding affective support for each other (Anderson, in press). The political nature of language use in online learning communities is, however, under explored.

Online educational communities typically use computer-mediated communication in its various forms as the primary technology for communication, although there is no doubt that such communities often do use a range of those forms and include other technologies such as phone and fax (Haythornthwaite *et al.*, 2000). A recent report (US Department of Education National Center for Education Statistics, 2003) indicates the level of use, and the extent of the preference for the use of asynchronous technologies at college level in the United States. Across all institutions offering distance education courses, 90 per cent use asynchronous computer-based technologies as the primary mode of instructional delivery, and 43 per cent use synchronous technologies. The primary means of communication for online communities is through text-based asynchronous computer-mediated communication and it is increasingly used within distance education courses and on-campus.

Democracy Online?

Text-based asynchronous computer-mediated communication (CMC will refer to this) enables group discussion and allows communication between individuals within an online course. It affords frequent and reasonably rapid dialogical possibilities. Harasim (1987; 1995) strongly advocated that these features contributed to the creation of a democratic environment for students because of the way learner participation is encouraged. In Harasim's view, inclusion is fostered because online courses expand access to educational opportunities, and the inherent nature of the medium allows all voices to be heard, and eliminates, or at least disguises, socially differentiating factors such as gender, physical disability or appearance. Students, she said could 'become "power learners" (Davie & Wells, 1991), taking control of their own education and playing an active and meaningful role in courses' (p. 218).

There is support for this view, especially for the argument that learners can open up discussion and take control of their learning. Rohfeld & Hiemstra (1995) have mentioned the role that computer-mediated communication can play in helping learners take control, Graddol (1989) noted the capability of computer-mediated communication to support minority topics of discussion and Tuckey (1993) noted the equalising effect of computer-mediated communication such that 'learners interact without regard to the status of other participants' (p. 64).

This portrayal of the computer-mediated communication environment in education paralleled a similar portrayal in the wider computer-mediated communication arena. Early writing provided a picture of this world as one where freedom, diversity and equality were the natural order. In both cases, the attributes of the medium are often cited as the basis of this argument. The relative anonymity of the text-based electronic medium, it was argued, would lead to a lessening of inequality in communication. This would arise because the lack of face-to-face presence would diminish the impact of social cues such as those associated with race, gender and organisational status. For example, in an early history of the Internet, Stirling (n.d.) wrote that one of the main reasons people wanted to be on the Internet was simple freedom, with no social or political protocols to hamper communication. This view of the Internet as enabling an open, equal democracy also underlies the writing of Rheingold (1993), who discussed the friendship and strong personal relationships that can be developed within online environments.

There is, however, compelling argument that online communities are political spaces where power and control are integral features of discussion. In the following sections of this chapter, I will provide a brief argument that points to the power-based nature of language use in face-to-face educational contexts, provide a more extended argument that language use in CMC is similarly constructed and, then, to illustrate this using race and gender as examples, discuss research that shows how online communication is not automatically democratic or egalitarian. From this general line of argument, I will move to a discussion related to more specifically educational contexts. Here I will draw on studies to illustrate the ways in which inequalities in online interaction are evidenced, but move beyond those points to suggest several ways in which educational interaction can usefully be structured and developed to benefit all participants.

Language and Power in Face-to-Face Contexts

In the context of this chapter, the comments of Barritt (1998) are particularly noteworthy. 'In one sense, CMC-based education is different from conventional education and needs to be approached as a new context, not as a simple change in mechanism. But in another sense the issues highlighted here are the same issues conventional education has struggled with for centuries; communication, interaction and cohort formation, productive control and authority relationships, and responsibility for learning' (para.1). A brief discussion of teaching and learning in conventional settings, in which the focus is on the role of language in education, is therefore warranted.

Mercer (1995) wrote about a process 'in which one person helps another to develop their knowledge and understanding. It is at the heart of what we call "education" (though education involves much more) and it combines both "teaching" and "learning" ' (p. 1). The process, that he called the guided construction of knowledge, is based around talk between teachers and learners. Mercer's work is based on work by Edwards and Mercer (1987). That earlier research was about 'the ways in which knowledge ... is presented, received, shared, controlled, negotiated, understood and misunderstood by teachers and children in the classroom' (p. 1). The social and language-based nature of the teaching-learning process was explored extensively by Edwards and Mercer (1987); and by Mercer (1995). Their work is based very strongly on a Vygotskian perspective of learning (Vygotsky, 1978; Wertsch, 1985). Although Vygotsky's work, and that of Edwards and Mercer, is based upon research with children, there is evidence that this approach can be applied to adult learning and development as well (Bolton & Unwin, 1995; Gallimore & Thorpe, 1990; Moll, 1990).

A theme emerging from the study of the language-based nature of learning is the role of power and control since 'education is necessarily ideological and predicated upon social relations in which power and control figure largely' (Edwards & Mercer, 1987, p. 161). This theme of power and control in educational discourse is echoed in Edwards and Westgate's (1994) work on classroom talk and is also evident in adult education. For example, Johnson-Bailey and Cevero (1998) and Tisdell (1993) demonstrated how power relationships are played out in adult education settings, largely within the verbal interactions between participants. Where these examples relate to the actual language in use, what Fairclough refers to as the power *in* discourse, Fairclough (1989) goes beyond the power of language itself to talk about the ways in which education also controls access to particular types of discourse, describing this as one of the aspects of power *behind* discourse (p. 62). As an example, Wegerif's work (1998) will be discussed in a later section to illustrate the effect of power behind discourse within CMC environments.

This brief overview provides the basis for the well-accepted view that power is closely implicated in the use of language in face-to-face educational contexts. While a similar perspective on the nature of language use in CMC is more contentious, there is considerable evidence to support such a perspective.

Accounting for Power and Control in CMC

The view of the use of CMC described previously, setting out a 'democratic' model of CMC use, is still fairly widely accepted. However, there are oppositional views about the nature of asynchronous online discussions and, with increasing awareness of the nature of online interaction, the early views are being modified. Two examples illustrate this re-working.

In work with a group of students involved in a LOTE class, Knobel *et al.* (1998) studied interactions between students using email for classwork in an on-campus graduate class for language teachers. The results of their study show that there should not be automatic support for several claims for the use of electronic communication relating to enhancing student autonomy, facilitating higher quality discussion and enhancing learning skills. More importantly though, Knobel *et al.* (1998) acknowledged that the most significant influence on learning outcomes is the particular purpose for which the networks are used. Their study is useful because it stresses that it is the situation of use that determines how, in what form, to what extent, and with what outcomes, interaction will occur.

In an early study, Janangelo (1991) wrote about the fiction that technology would bring better teaching and implicit equity. In writing about the use of CMC in classrooms, his concern was about the way technologies have been used to support oppressive social relations (technopower) and how technopression, the use of technology to oppress or to ensure self-subjugation, can occur within educational contexts. This is strongly at odds with the assertion that CMC will provide a more democratic education environment for discussion in which people are on equal terms as they contribute, take turns and raise topics.

These two examples provide the beginning of a different story of CMC use. The earlier work of theorists in the area of computer-mediated communication was perhaps characterised by a sense of technological determinism that more recent work is shrugging off. Jones (1997) provided a full argument to support the rejection of technological determinism saying, 'the particular form that an individual virtual community takes is not determined by technology but rather is dependent on its social context' (p. 10). That argument is elaborated in a telling critique of the 'democratic' model of CMC use offered by Spears and Lea (1994) in their discussion of power in computer-mediated communication environments.

Spears and Lea (1994) started by giving careful consideration to the ways in which power is being characterised. Control, they said, can be characterised as control over one's work so that power is then defined as the ability to get things done. They offer the critique of this view by saying: 'However, the narrowness of this approach becomes clear when we consider that neither power nor control so defined reflect an increase in the social or relational power of the subordinate … still less a release from the power relationship. Expansion of individual control over one's work domain or productivity by virtue of technological forces is conceptually distinct from the social power relations within which this is exercised' (p. 435).

At this level of analysis, the supporters of electronic democracy could argue that they acknowledge the nature of power relations that exist between teacher and students in classrooms of all sorts (although few do), and are prepared to accept those relationships as part of education. Perhaps it is the democratising of relationships between students that is important. Rohfeld and Hiemstra (1995) seemed to reflect this subtlety in their discussion of students who described themselves as 'timid, passive or unable to think quickly … in face-to-face situations' (p. 102). These students stated that CMC gave them time to reflect and compose class discussion contributions, and the authors concluded that CMC was valuable, in part, because 'helping learners take increasing control over personal learning is a goal for most educational endeavors' (p. 102).

Further analysis by Spears & Lea (1994) made even this assertion problematic. They start by acknowledging the ways in which power has been traditionally conceptualised and discussed as an 'external force to which the individual succumbs; power is typically imposed from above, against the will of the resisting and reluctant individual' (p. 436). They go further than this rather direct sense of power and argue that even the portrayal of social pressure suggests an external normalising force. They write that social pressure leads to 'compliance or submission to the views of powerful others, or those with the power to reward or provide such approval. Such compliance is virtually indistinguishable from the effects of power per se' (p. 436).

Having set out this view of power as an external, direct force they turn to a more subtle rendering of the nature of power drawing on a Foucauldian analysis. Spears and Lea (1994) suggested that: 'the effects of both power and influence rely on an active agent to exert their effects; they could not exert their effects on inert objects, but rely on the identity and agency of both parties and the social relations that bind them. ... It follows that if power and influence are not outside, but are at least partly encoded with us, it becomes far less easy to argue that the source of power is necessarily displaced or diluted by ... CMC' (p. 437)

This argument represents a well-reasoned rebuttal of the assumption of an inherently democratic communicative environment for CMC-based communities. Indeed, Herring (1996a) reported on this earlier work, noting the 'Utopian visions of class- and gender-free virtual societies' (p. 1) that formed part of the work of some earlier theorising. More recent work takes the view that the ideal of an inclusive or egalitarian environment is not easily established and that students taking control of their learning or having their voices heard is not something to be achieved in a straightforward manner. This work confirms the need to see CMC use as a complex social phenomenon, especially with regard to the concepts of control and power, concepts that are generally noticeable through their absence in discussion of online interaction in education.

Jones (1998) was very clear about the importance of considering the issue of power in online discussions. Noting that the idea of communities emerges strongly in the wider literature on computer-mediated communication, he wrote that 'communities are defined not as places but as social networks, a definition useful for the study of community in cyberspace ... [since] ... it focuses on the interactions that create communities' (p. 20). He went on to say that 'just because the spaces with which we are now concerned are electronic there is not a guarantee that they are democratic, egalitarian or accessible and it is not the case that we can forgo asking in particular about substance and dominance' (p. 20).

While many studies have concentrated on the immediate situation of use, it is possible to look more widely at the structures and relationships that are carried into the online environment. Kramarae (1998) wrote that 'Cyberspace can provide freedoms of various sorts, but they are designed and constrained by powerful structured forces of assumptions and goals; they are not equally friendly environments or opportunities for everyone' (p. 113). Also looking beyond the immediate, Erickson (1997) noted that properties of the medium that encourage particular communicative features he saw in his analysis of online discourse could easily support alternative features. He suggested that what prevents the alternatives from occurring are primarily social factors (the nature of the online 'community') and institutional factors such as the policies related to managing the discourse. Baym (1995) is another who looked more widely for forces that might shape discussion, and noted the need to consider the structural features of the tools of use as resources used to create an online culture. Although technological deterministic approaches can be rejected, it does not do to reject the notion that technology has an effect on communication.

These more general studies have highlighted the idea that technologically determinist views of how discussion might be undertaken in an inclusive or egalitarian way must be discounted. In addition, they have shown that although social forces underpin the use of technology, all technologies also bring their own affordances to their use because of the social origins of their design. Those affordances will have an impact on the way technologies are used. These studies have also shown that in the immediate situation of use, the dynamics of power relations are still likely to exist and those relations should be taken into account in any consideration of the relationship between control and online interaction.

Gender, Race and CMC

Issues surrounding race and gender have been investigated in studies of computer-mediated communication contexts. These studies tend to show that gender- and race-based differences do exist and are visible online. In this section the areas of gender and race are considered more fully.

Gender

Herring's work provides the most useful empirical analysis of the impact of gender in CMC (Herring, 1993, 1996b, 2000). Herring conducted most of her work using publicly accessible mailing lists and Usenet mailing lists and has also drawn widely on a range of studies of gender and CMC in her writing. She suggests that the number of findings of gender disparity in the nature and extent of CMC use 'raise an apparent paradox: how can gender disparity persist in an anonymous medium which allegedly renders gender invisible?' (Herring, 2000, p. 3). In responding to this apparent paradox she has focused on asking both 'what happens?' and 'why?'

Gender-based differences in levels of participation are evident in the two lists (LINGUIST and MBU) studied by Herring (1993). Although women were list subscribers in reasonably high numbers (42 per cent on MBU and 36 per cent on LINGUIST), they contributed far less than men. A difference in the level of participation was also discernibly topic related. For example, sexism-related topics drew participation from 30 per cent of the women, while theory-related discussions drew input from only 16 per cent. In addition, messages from women were typically much shorter than those from men.

A second area addressed by Herring (1993) concerns gender characteristics in online interactions, with a comparison between the language used by women and men. An analysis of the language features used in online communication by men and women was based on the following set of features hypothesised to characterise a gendered stylistic variety.

Table 1: Language features used in online communication by gender.

Women's language	Men's language
Attenuated assertions	Strong assertions
Apologies	Self-promotion
Explicit justifications	Presuppositions
Questions	Rhetorical questions
Personal orientation	Authoritative orientation
Support of others	Challenge to others
	Humour/sarcasm

Herring found large differences in the style of language use. While the (male or female) style of language use is not the exclusive domain of males or females, Herring noted that 'gender predicts certain online behaviours with greater than chance frequency when considered over aggregate populations of users' (Herring, 2000, p. 4). In particular, she found that in the lists she studied, men tended to use the male discursive style and women employed a style that was a mixture of both male and female styles, and concluded that it was easier for men to maintain a distinct style (masculine, feminine, or neutral).

Herring also reports on what she calls the 'stereotype of the informational male and the interactive female computer user' (Herring, 1996, p. 81). Yates (2001) discusses the links between this stereotype and the boundary between 'legitimate' informational content, and 'illegitimate' personal content. Herring suggests, as does Yates, that the stereotype and thus the boundary demarcating acceptable from unacceptable content is based on a sociocultural background that values informational exchange, and where 'men are expected to be knowledgeable, rational and dispassionate' (p. 105). However Herring's investigation into this area allowed a more subtle rendering of this issue. She found that men and women post messages that are structured in interactive ways and that the exchange of views takes precedence over the exchange of information. While both purposes are evident in the messages of both gender groups, Herring also found that 'women and men negotiate information exchange and social interaction in gendered ways' (p. 82). Where women's messages are aligned and supportive, men's messages tend to be critical and oppositional. Of interest is the additional finding that within the lists in which the discussions were occurring, there were tendencies to alignment or oppositional messages depending on the relative number of each gender. In other words, the (gender) minority members of each list tended to shift their style of communication toward that of the (gender) majority.

Herring's work, supported by the findings of many others (e.g. Baym, 1996; Hall, 1996; Hert, 1997; Savicki *et al.*, 1996) shows that gender is far from invisible in online discussions; and the gendered nature of discussion has ramifications for the structuring and control of CMC (see also Murray, 2000). Yates (2000) summarises these ramifications by saying 'The "democratic" model has not won out and, as with face-to-face education situations, gender has a key role to play in structuring the interactions so as to marginalize women's contributions' (p. 27).

Race

Researchers into race in cyberspace reach largely similar findings as those within the area of gender research. The conclusion offered by Kolko *et al.* (2000) is that:

> You may be able to go online and not have anyone know your race or gender – you may even be able to take cyberspace's potential for anonymity a step further and masquerade as a race or gender that doesn't reflect the real, offline you – but neither the invisibility nor the mutability of online identity make it possible for you to escape your 'real world' identity completely. Consequently, race matters in cyberspace precisely because all of us who spend time online are already shaped by the ways in which race matters offline, and we can't help but bring our own knowledge, experiences, and values with us when we log on (pp. 4–5).

This conclusion comes from the introduction to a collection of work that describes the various ways in which race is presented online and, in other circumstances, pointedly ignored because to acknowledge its presence would be disruptive. In relation to this last point, the authors in this collection argue that often in the spaces within which Internet interaction occurs, the race switch is kept in the 'off' position or at very least is kept in a default 'White' position (Lockard, 2000).

When the race switch is 'on', when questions of race are a focus, the same matters of language style and use emerge as identifiers. Using Usenet groups, Burkhalter (1998) demonstrated how racial identity is made visible online. He described how self-disclosure and the use of language serve to provide ways in which one's racial identity is disclosed. Burkhalter suggests further that identity is negotiated through interaction and not the exclusive claim of a person. He says that those engaged in 'Online interaction (use) an individual's perspectives, beliefs and attitudes to make assumptions about the individual's racial identity' (p. 62) and contrasts this with the offline case, where a person's (physical) racial identity might lead to assumptions about beliefs, attitudes, and perspectives. What is clear from his work is that racial identity is clearly signalled online, and that the racial world online resembles its offline counterpart.

These examples for race and gender show clearly that interaction on the Internet is not necessarily the free, open, and unhindered passage of ideas and opinions that a democratic model might lead us to expect. The words of the MCI television commercial, 'There is no race. There is no gender. There is no age. There are no infirmities. There are only minds. Utopia? No, Internet' have a hollow ring to them.

CMC in Education
Interaction and control
The previous section discussed the ways in which issues related to gender and race are enacted in CMC generally. While the literature concerning gender and race in online interaction is not extensive, it is sufficient to provide convincing evidence that language use in online contexts is not devoid of the same social and political forces that shape face-to-face communication. Within online education there has been even less emphasis on the exploration of these issues, despite the larger encompassing field of distance education having a reasonable literature base of gender-based research developed through the 1980s and early 1990s. However, as the following paragraphs will show, there should be no less awareness and understanding of the ways in which control is exerted over the nature and extent of CMC within online learning communities. This small collection of studies provides an illustration of the varied ways in which control is encountered and then shapes the interaction that occurs online.

Where studies focused on Usenet are likely to be discussing large, open, very flexible groupings of people, those within an education context feature smaller, stable groups of students. Within this more tightly constrained context, group development is an important issue. McDonald and Gibson (1998) investigated the question of how learning groups are formed, maintained, nurtured and developed. They coded a sample of messages from transcripts selected at three points in an online course and found that there is a definite pattern to interpersonal issues in group development. They explored development in terms of the interpersonal dimensions of inclusion (involvement), control and affection (openness and solidarity) based on coded analysis of transcripts of group discussion, and found that group development moves through stages that can be identified and can be negotiated to form a cohesive functioning group. McDonald and Gibson found concern over control and involvement decreased over the life of the group (thirteen weeks) while solidarity and openness increased, as was hypothesised. This decline in concern with control does not necessarily imply the absence of power or control relations within the group. An alternative explanation is that power relations have been cemented and accepted – there is no guarantee of group democracy.

The difficulty of developing a sense of equitable democratic involvement in group interaction is also seen in Wegerif's (1998) work. Wegerif discussed the formation and growth of an online educational community in his analysis of a CMC-based British Open University course. He talked of students crossing a threshold that gave them full participation status within their learning community. Crossing this threshold was contingent on factors of access (cost and technical elements), gender, and prior experience with group work, and was an important element of feeling and being successful within the course. Students noted the formation of an 'in-group' within the course and Wegerif identified the need to move students from 'outsiders' to 'insiders'. Those who do not make the transition often do not complete the course.

The role of prior experience in CMC, one of Wegerif's threshold factors, was also noted by Vrasidas and McIsaac (1999) in their examination of the nature of interaction in an online course. They found that the structure of the course, class size, feedback and prior experience with computer-mediated communication all influenced interaction. The study suggested that educators can provide structure to ensure online interaction and that learner–learner interaction was an important part of that interaction. Their concern about the impact of previous experience and the intimidation felt by those with less prior experience led them to suggest the need to explore power and control in online groups. In this regard they said that a 'discourse analysis approach would shed light on how the ideas of power and control operate in online and face-to-face encounters. It would be useful to explore what discourse says about power and to examine how interaction shapes power in an online environment' (p. 34).

The style of messaging is also a concern for some students. Wegerif noted 'Considerable concern and anxiety about the form of messages was evident' (1998, p. 40). One student found the gender bias she experienced in work situations was not present in the online group. However, some students posted long messages which others found intimidating. The form of messages and the style of interaction (such as argumentation or 'cumulation') all impacted on the extent to which CMC supported an egalitarian style of communication. In a totally different context, that of Japanese speaking (and writing) professionals in the field of teaching English to speakers of other languages, Matsuda (2002) reports a similar phenomenon. Through the analysis of features of the Japanese language, the study shows that the social relations between members of an online community slowly shifted to be based on the extent of knowledge members demonstrated through their online discourse. This knowledge display through interaction became a critical factor in the development of hierarchical relations.

Although limited experience in CMC use might lead to feelings of intimidation, experience is no guarantee of communicative ease. In an evaluation study of an online class, Picciano (1998) found that, in a course with experienced students, concern about speaking in class was still evident: 'While some students prefer speaking up in class, others do not' (Picciano, 1998, p. 11). The asynchronous environment did not remove that anxiety, it reshaped it. Students recognised that their responses were open to more scrutiny because they were available for continual review. However, students did like having more time to draft their responses.

The author also found there were some more democratic aspects. Based on counts of responses, Picciano wrote, 'Student and instructor roles were changed in the asynchronous course. The students had more of a voice in the discussions' (1998, p. 11). While message numbers do suggest some measure of democratic input to discussion, the claim here does not account for the weighting or authority assigned to those voices that are not necessarily equal. To claim, as Picciano did, that the class was therefore accepting empowerment and responsibility is unwarranted.

A study by Burge (1994) shows how the roles of students and instructors are differently constructed. Burge's study involved Masters of Education students who were using CMC. The study identified four styles of peer behavior that were required in online interaction: participation, response, provision of affective feedback and short focused messaging. Two key behaviours for instructors were required: discussion management (structure, pacing, focusing, setting protocols) and contribution (technical help, timely individualised content-related messages and feedback, summaries, offering support). There was a clear difference in the roles expected of instructors and students. The idea of management and the expectation of expertise suggest authority vested in some group members – in this case, the instructors.

The role of the teacher noted in Burge's work is also evident in the work of Hillman (1999). Hillman provided analysis of discourse taken from four face-to-face courses and two taught via CMC using an analytic technique that, it is claimed, combines both approaches. The discourse is analysed in terms of purpose, mechanism and content. Hillman's analysis demonstrated that instructors in the CMC environment took a clearly defined role – they spoke more often and at greater length than students.

These studies, even with their focus on directly educational rather than social variables, still demonstrate that the issue of control in online interaction must be of concern to educators. The spaces within which online learning communities interact are, as Jones said, not guaranteed to be egalitarian or democratic and are subject to issues of substance and power.

Moving Forward

The issues concerning gender and race in CMC interactions, alongside those factors mentioned in the previous section, all reflect a social situation rather than the specific features and affordances of CMC. In this regard, online interaction is no different from the face-to-face interaction that routinely occurs in the practice of education at all levels. Being aware, and taking advantage of the socially grounded nature of interaction provides the basis from which educators can act to ensure that interaction in online learning communities is enabling for the learning of all students, not just some.

The appropriate social organisation of a CMC system can reduce the effects of hierarchical social relations. As Collins-Jarvis (1993) notes in relation to gender and Silver (2000) in relation to race, providing or enabling the invention of spaces in online learning communities within which specific groups can develop their own communicative and social goals and discuss their own issues is central to maintaining involvement. These spaces provide sanctuary at times, but such authors (see also Camp, 1996, Zickmund, 1997, cited in Silver 2000) also note the importance of engagement in the larger online community.

In comparison with the macro considerations involved in creating the structures for communication in online learning communities, there are actions that can be taken by individual educators. On the basis of their experience within a bicultural online environment, Sujo de Montes *et al.* (2002) note three such actions. First is the need to teach students to critically reflect on their own words and interactions online:

expecting such reflection of students is conditional on the construction of online environments within which students can safely examine the relationship between their personal history and their online behaviours. Therefore, second, instructors must develop the skills that enable them 'to make explicit what is implicit in people's words, actions, and expectations' (p. 269) and, third, instructors must be 'willing to analyze their own biases and assumptions, first when they build online courses and then when they interact with online students' (p. 269).

These considerations, both macro- and micro-, have not yet become part of the rhetoric of online education. In this chapter, I have presented an argument that suggests why they must be routinely discussed and enacted among those who create and interact within online learning communities. I have shown that there is no room for the 'democratic' model of CMC within discussion of online interaction. But although there are no grounds for proclaiming the inherent democracy of CMC, there is no suggestion that educators cannot legitimately work towards a goal of supporting the educational needs and goals of all students. Yates (2001) wrote of CMC that only 'once the technology is socially and educationally organised in ways which (students) can adapt to their own norms [can it] provide a very useful basis for educational interaction' (p. 33). We all work towards that time.

References

Anderson, B. (in press). Dimensions of learning and support in an online community: A case study. *Open Learning*.

Barritt, M.D. (1998). *Extending educational communication: the experience of distance learning via a distributive environment for collaboration and learning (computer mediated communication).* Unpublished doctoral dissertation for the *University of Michigan*. Retrieved 14 March 2000 from: Proquest Digital Dissertations AAT 9825166

Baym, N.K. (1995). From practice to culture on Usenet. In S.L. Star (ed.), *The cultures of computing*. Oxford: Blackwell.

Baym, N.K. (1996). Agreements and disagreements in a computer-mediated discussion. *Research on Language and Social Interaction, 29*(4), 315–45.

Bolton, N. & Unwin, L. (1995). Apprenticeship in distance learning. *Open Praxis, 2*, 11–13.

Burge, E.J. (1994). Learning in computer conferenced contexts: The learners' perspective. *Journal of Distance Education, 9*(1), 19–43.

Burkhalter, B. (1998). Reading race online: Discovering racial identity in Usenet discussions. In M. Smith & P. Kollock (eds), *Communities in cyberspace*. New York: Routledge.

Collins-Jarvis, L.A. (1993). Gender representation in an electronic city hall: Female adoption of Santa Monica's PEN System. *Journal of Broadcasting and Electronic Media, 37*(1), 47–66.

Edwards, A. & Westgate, D.P.G. (1994). *Investigating classroom talk* (2nd edn). London: Falmer Press.

Edwards, D. & Mercer, N. (1987). *Common knowledge: The development of understanding in the classroom*. London: Methuen.

Erickson, T. (1997). *Social interaction on the net: Virtual community as participatory genre*. Paper presented at the Thirtieth Hawaii International Conference on Systems Science, Hawaii.

Fairclough, N. (1989). *Language and power*. London: Longman.

Gallimore, R. & Thorpe, R. (1990). Teaching mind in society; teaching schooling and literate discourse. In L. C. Moll (ed.), *Vygotsky and education: instructional implications and applications of sociohistorical psychology*. Cambridge: Cambridge University Press.

Gee, J.P. (1999). *An introduction to discourse analysis*. London: Routledge.

Graddol, D. (1989). Some CMC discourse properties and their educational significance. In R. Mason & A. Kaye (eds), *Mindweave: Communication, computers and distance education*. Oxford: Pergamon.

Hall, K. (1996). Cyberfeminism. In S.C. Herring (ed.), *Computer-mediated communication: Linguistic, social and cross-cultural perspectives* Philadelphia: John Benjamins, 147–170).

Harasim, L. (1987). Teaching and learning on-line: Issues in computer-mediated graduate courses. *Canadian Journal of Educational Communication, 16*(2), 117–135.

Harasim, L., Hiltz, S.R., Teles, L. & Turoff, M. (1995). *Learning networks: A field guide to teaching and learning online*. London: MIT Press.

Haythornthwaite, C., Kazmer, M. M., Robins, J. & Shoemaker, S. (2000). Community development among distance learners: Temporal and technological dimensions. *Journal of Computer-Mediated Communication, 16*(1). Retrieved on 22 November 2002, from <http://www.ascusc.org/jcmc/vol16/issue1/haythornthwaite.htm>

Herring, S.C. (1993). Gender and democracy in computer mediated communication. *Electronic Journal of Communication, 3*(2).

Herring, S.C. (1996a). *Computer-mediated communication: Linguistic, social and cross-cultural perspectives*. Philadelphia: John Benjamins.

Herring, S.C. (1996b). Two variants of an electronic message schema. In S.C. Herring (ed.), *Computer-mediated communication: Linguistic, social and cross-cultural perspectives*. Philadelphia: John Benjamins.

Herring, S.C. (2000, January 7, 2004). Gender differences in CMC: Findings and implications. *The CPSR Newsletter, 18(1), 3-11*. Retrieved 7 January 2004, from: <http://www.cpsr.org/publications/newsletters/issues/2000/Winter2000/herring.pdf>

Hert, P. (1997). Social dynamics of an on-line scholarly debate. *The Information Society, 13*, 329–60.

Hillman, D.C.A. (1999). A new method for analyzing patterns of interaction. *The American Journal of Distance Education, 13*(2), 37–47.

Janangelo, J. (1991). Technopower and technopression: Some abuses of power and control in computer-assisted writing environments. *Computers and Composition, 9*(1), 47–63.

Johnson-Bailey, J. & Cevero, R.M. (1998). Power dynamics in teaching and learning practices: An examination of two adult education classrooms. *International Journal of Lifelong Education, 17*(6), 389–99.

Jones, Q. (1997). Virtual communities, virtual settlements and cyber-archaeology: A theoretical outline. *Journal of Computer Mediated Communication, 3*(3). Retrieved 16 January 2001, from: <http:www.ascusc.org/cmc/vol3/issue3/ones.html>

Jones, S.G. (1998). *Cybersociety 2.0: Computer-mediated communication and community*. Thousand Oaks, CA: Sage.

Knobel, M., Lankshear, C., Honan, E. & Crawford, J. (1998). The wired world of second language education. In I. Snyder (ed.), *Page to screen: Taking literacy into the electronic era*. New York: Routledge.

Kolko, B.E., Nakamura, L. & Rodman, G.B. (2000). Race in cyberspace: An introduction. In B. E. Kolko, L. Nakamura & G.B. Rodman (eds), *Race in cyberspace*. New York: Routledge, 1–14.

Kramarae, C. (1998). Feminist fictions of future technology. In S.G. Jones (ed.), *Cybersociety 2.0: Computer-mediated communication and community*. Thousand Oaks, CA: Sage.

Lockard, J. (2000). Babel machines and electronic universalism. In B. E. Kolko, L. Nakamura & G.B. Rodman (eds), *Race in cyberspace*. New York: Routledge.

Matsuda, P.K. (2002). Negotiation of identity and power in a Japanese online discourse community. *Computers and Composition, 19*(1), 39–55.

McDonald, J. & Gibson, C.C. (1998). Interpersonal dynamics and group development in computer conferencing. *American Journal of Distance Education, 12*(1), 7–25.

Mercer, N. (1995). *The guided construction of knowledge*. Clevedon: Multilingual Matters.

Moll, L.C. (1990). *Vygotsky and education: Instructional implications and applications of sociohistorical psychology*. Cambridge: Cambridge University Press.

Murray, D. (2000). Protean communication: The language of computer-mediated communication. *TESOL Quarterly, 34*(3), 397–421.

Picciano, A.G. (1998). Developing an asynchronous course model at a large urban university. *Journal of Asynchronous Learning Networks 2*(1). Retrieved 13 March 2000, from: <http://www.aln.org/alnweb/journal/jaln_vol2issue1.htm#picciano>

Rheingold, H. (1993). *The virtual community.* Retrieved 12 January 2001, from: <http://www.december.com/cmc/mag/1997/jan/toc.html>

Rohfeld, R. W. & Hiemstra, R. (1995). Moderating discussions in the electronic classroom. In Z.L. Berge & M. P. Collins (eds), *Computer mediated communication and the online classroom.* Cresskill, NJ: Hampton Press, 11–24

Savicki, V., Lingenfelter, D. & Kelley, M. (1996). Gender language style and group composition in Internet discussion groups. *Journal of Computer-Mediated Communication, 2*(3). Retrieved 19 December 2003, from: <http://www.ascusc.org/jcmc/vol2/issue3/savicki.html>

Silver, D. (2000). Margins in the wires: Looking for race gender and sexuality in the Blacksburg Electronic Village. In B.E. Kolko, L. Nakamura & G.B. Rodman (eds), *Race in cyberspace.* New York: Routledge, 1–14

Smith, C.B., McLaughlin, M.L. & Osborne, K.K. (1996). Conduct control on Usenet. *Journal of Computer-Mediated Communication, 2*(4). Retrieved 28 February 2003, from <http://www.ascusc.org/jcmc/vol2/issue4/smith.html>

Spears, R., & Lea, M. (1994). Panacea or panopticon: The hidden power in computer mediated communication. *Communication Research, 21*(4), 427–59.

Stirling, B. (n.d.). *History of the Internet.* Retrieved 19 January 2001, from: <http://www.dsv.su.se/internet/documents/internet-history.html>

Sujo de Montes, L.E., Oran, S.M. & Willis, E.M. (2002). Power, language, and identity: Voices from an online course. *Computers and Composition, 19*(3), 251–71.

Tisdell, E.J. (1993). Interlocking systems of power, privilege and oppression in adult higher education classrooms. *Adult Education Quarterly, 43*, 203–26.

Tuckey, C.J. (1993). Computer conferencing and the electronic whiteboard in the United Kingdom: A comparative analysis. *American Journal of Distance Education, 7*(2), 58–72.

US Department of Education National Center for Education Statistics. (2003). *Distance education at degree-granting postsecondary institutions: 2000–2001* (No. NCES 2003 - 017). Washington DC: US Department of Education.

Vrasidas, C. & McIsaac, M.S. (1999). Factors influencing interaction in an online course. *American Journal of Distance Education, 13*(3), 22–36.

Vygotsky, L. (1978). *Mind in society.* Cambridge MA: Harvard University Press.

Wegerif, R. (1998). The social dimension of asynchronous learning networks. *Journal of Asynchronous Learning Networks, 2(1).* Retrieved 13March 2000, from: <http://www.aln.org/alnweb/journal/jaln_vol2issue1.htm#wegerif>

Wertsch, J. (ed.) (1985). *Culture, communication and cognition: Vygotskian perspectives.* New York: Cambridge University Press.

Yates, S. J. (2001). Gender, language and CMC for education. *Learning and Instruction, 11*, 21–34.

Creating Online Professional Learning Communities: A Case of Cart before Horses

8

Judy M Parr and Lorrae Ward

One of the ways that education has taken up the possibilities of electronic communication is to create and support online communities of educational professionals. The use of computer-mediated communication is seen as a valid form of professional dialogue, support and exchange. Promoting professional online communication is often a major part of the attempt to make the use of ICT part of day-to-day practice in schools. However, the suggestion is that much 'hyperbole' surrounds online forums (Selwyn, 2000, p. 751) and there is a need for research to examine more closely the many claims.

FarNet: Learning Communities in the Far North

FarNet is one of four Digital Opportunities pilot projects funded by the New Zealand Ministry of Education in partnership with business. The FarNet project in the relatively isolated northern region of New Zealand is an example of an effort to establish a linking of teachers in ten schools through the use of the web. Schools were already part of an ICTPD Cluster. As part of Digital Opportunities, FarNet schools received hardware, software and broadband access. The partnership protocol, between government and business, identifies the project by the term 'Learning communities in the Far North' and indicates that a major goal of the project was 'to support changes in access and attitudes to learning as well as a culture of collaboration across schools', the latter particularly in terms of 'curriculum planning and delivery'. In interviews associated with scoping the project, two of the three participating schools mentioned as success indicators for the project, sharing or changes in connectivity and sharing. As Hargreaves (1994) notes, collaboration and collegiality are widely seen as means of ensuring effective implementation of change introduced from outside. A major aim of FarNet was to provide the means by which teachers could produce and then share electronic resources. The implication in FarNet was that successful posting and sharing of resources and email communication by teachers would be a significant factor in the success of the introduction of Internet technology and, relatedly, ICT use in the schools.

Learning in Order to Change Practice

The aim of establishing learning communities was linked to another aim of the FarNet project, namely, to bring about a change in teacher pedagogy and to enhance student learning outcomes. We will not enter here into a debate about whether ICT is a successful change agent or, as is more likely, a lever in terms of pedagogical change (Venezky & Davis, 2002). Nor will we debate the fraught question of whether ICT has been shown to enhance student learning outcomes or tackle the issue of different pedagogies leading to differential outcomes (see Becker, 2001; Cox *et al.*, 2004a,

Creating Online Professional Learning Communities 125

2004b; Kulik, 1994; Niemic & Walberg, 1987; Shakeshaft, 1999). Suffice to say that assumptions that ICT would both effect a change in pedagogy and lead to enhanced achievement on the part of students were held by the proponents of the project.

Rather, we will consider the extent to which the FarNet project supported the creation of a professional learning community focused on enhanced teacher practice and student outcomes. The project largely attempted to do this through the development of a website with associated pages for curriculum areas and special interest groups. There were also other areas such as a student page and a listserv for each curriculum area. In considering the electronic community, we accept the notion of the largely theoretical work concerning the likelihood of change through the creation and operation of such communities (Little, 2000). We intend to draw on our research within this project, in which we used multiple methods to collect data to evaluate its outcomes. These methods included interviews with key personnel such as ICTPD coordinators in each school and teachers who had taken up roles as 'leaders' of their curriculum area within FarNet, as well as self-report from questionnaires to all teachers, largely concerning their level of skill and confidence in using ICT, their level and type of use and more specifically their use of both the Internet and the FarNet site. Also, we monitored the FarNet site and associated listservs over time and considered available evidence of enhanced student learning. We aim to illustrate some of the issues and challenges surrounding virtual professional interaction and learning and also to suggest the nature of the interlocking pieces in the development of an effective model for an online community.

The Notion of Community

The word community (and, to some extent, online community) 'has become an obligatory appendage to every educational innovation' (Grossman *et al.*, 2001, p. 942). The word community is used in various ways in relation to teachers and their professional work. Common examples are teacher community, school community and community of practice. It is far from clear if there are common features across such use. In terms of how a community might function theoretically, Westheimer (1998) identified from the literature, common themes including interdependence, interaction/ participation, meaningful relationships, shared interests and concern for all views, but noted that research had yet to establish empirically the dimensions of the concept community. This does not prevent, in research, the liberal use of the term, almost to the point that a community is brought into being largely by 'linguistic fiat' (Grossman *et al.*, 2001, p. 943).

A Professional Learning Community

The rationale for professional communities of teachers is to provide an ongoing, sustainable vehicle for teacher learning (Cochran-Smith & Lytle, 1999; Darling-Hammond & Sykes, 2000). In theory, all learning is grounded in social interaction. Social interaction around learning can be conceived of as occurring on a continuum (Parr & Townsend, 2002) where, at one end, the interaction is largely one way like a tutor–tutee relationship, to the midpoint where learning is cooperative with all contributing towards a shared goal, to the other end of the continuum where learning is co-constructed. Various mechanisms are seen to contribute to learning. To explain to others or ask questions, one has to rehearse and be clear about what one knows. When new information that challenges existing ideas is encountered, an effort has to be made to reconcile the discrepancy and, often, new learning occurs as a result of restructuring. New learning may occur through the process of co-construction where the learning outcome may be more than the sum of the parts, the knowledge that each individual brings.

Professional learning communities have distinctive features that include: shared norms and values; collective learning through collaboration; the application of that learning in a focus on student learning, and shared personal practice and reflective dialogue (Kruse *et al.*, 1995). A genuine professional learning community does not involve 'comfortable collaboration' where the privacy of the teacher's classroom is protected and there is no deep probing of issues of teaching and learning (Fullan & Hargreaves, 1996). This is a 'weak' form of community (Little, 1990). Research suggests that strong professional learning communities are those focused on school change and improvement, engaged in what Little (1990) terms 'joint work'. Such work involves not only acquiring new knowledge (this centrally includes increasing teachers' pedagogical content knowledge – that is their knowledge of their subject, particularly from the point of view of how to teach it – as a significant part of enhancing practice) but also challenging and critiquing basic assumptions about teaching and learning. A professional community of teachers has, as its central aim, school-wide or even beyond-the-school efforts to improve practice and, as a consequence, student learning.

An irony is that, within teaching, much learning aimed at extending teachers' pedagogical content knowledge has taken place outside the workplace rather than within the logical venue, the school, so there is little history and few structures in place to support teacher learning *in situ* (Grossman *et al.*, 2001). Individual teachers vary in their degree of interaction both within and between schools and, given the large measure of autonomy associated with the profession, 'make individual choices on the basis of individual considerations within the context of the school organization'; teachers make decisions about the merit of interaction with colleagues by weighing up whether it detracts from their opportunities and time to work with their students (Bakkenes *et al.*, 1999, p. 167). There is no established culture of a collective responsibility for teacher learning. Traditionally, teachers' responsibility is to their own students rather than to other teachers or the students of those other teachers (Grossman *et al.*, 2001).

The Possibility of Strengthening Existing Communities

The potential of electronic communication is not necessarily realised for a variety of reasons. Using online technology to create networks (in the case of FarNet this was curriculum resource pages and email lists of teachers who taught each subject) might qualify as a case of putting the cart before the horse (Schlager & Fusco, 2003). This refers to the idea that such use to actually create networks for a specific purpose may be precipitous. Among other things, it often ignores the greater potential of the Internet to support and strengthen existing communities of practice.

Although FarNet attempted to utilise the idea of linking teachers who shared a common interest in that they taught the same curriculum area, such could hardly be said to constitute an existing community. Traditional curriculum-based associations such as the English Teachers Association or the Science Teachers Association could not claim to involve more than a small percentage of teachers. A curriculum leader interviewed felt that the FarNet structure should 'work in with our professional association [although] the association has disintegrated' but that FarNet might help in trying to get it started again. Within the FarNet schools, it was not common for teachers in one school to communicate with those in another. When asked to report on the level of communication they had had with colleagues from other schools prior to the implementation of FarNet, 45 per cent of respondents stated that they had had none, while just 2 per cent had done so on a regular weekly basis.

This idea that electronic communication might best build on existing communities was in part reinforced by the experience of at least one group within FarNet. FarNet's chosen organisational structure of curriculum groups, with leaders whose primary role was to facilitate the development and sharing of resources and also information and views about utilising technology in a particular subject area, implied a collective responsibility for teacher and student learning. But the curriculum leaders conceived of their role as little more than a conduit; they had little sense of what a professional learning community was or how it might function, and teacher development was not their responsibility. In reality, the member schools were often in competition with one another for students, because numbers are important in funding and retaining staff. So, such use of technology highlighted border politics issues (Achinstein, 2002). These two factors, namely, the notion of creation of web-based resources and electronic communication fitting more successfully within an existing community, and the idea of collective responsibility not sitting easily with the reality of competition, perhaps explain the differential success of groups.

The most successful group in terms of the extent of resources posted and the level of reported use was the Maori curriculum group. The Maori curriculum site at one point (July 2003) had 120 separate documents on it with thirteen links to other sites. This equates to about a third of the total resources and links posted within the nineteen curriculum group pages on the FarNet site at that point in time. Admittedly, there was a Maori resources coordinator who collected and posted resources; however, the point is that he still had to find people willing to give resources. Taking collective responsibility for children and their growth and development is a feature of Maori culture. So, Maori teachers felt responsibility for all Maori students, regardless of the schools they attended, or the subjects they took. They also felt collective responsibility for each other, not just as teachers of Maori but as Maori teachers in general. 'Maori teachers as teaching anything … I have got a Maori teacher who teaches science … And after the first year I found him helping with Te Reo Maori and that was neat …' In addition, Maori teachers in the region tended to have strong associations through tribal affiliations or family ties to one another and to many other Maori in the wider community beyond the school. These associations, together with the sense of community that characterises Maori culture, clearly helped Maori curriculum leaders to locate resources and persuade teachers to share them more widely by posting them.

A Case of the Cart before Several Horses

So, the cart before the horse analogy seems apt. Utilising and strengthening existing ties and affiliations appear to assist uptake in terms of participation in a virtual community. But, we argue that it is very informative to extend the cart before the horse analogy. In actual fact, in FarNet, the cart was placed before several horses that, if addressed and harnessed collectively, may have facilitated more effective outcomes.

Professional learning communities within a school horse

In this section we argue that experience of a well-functioning professional learning community within a school may predispose teachers to connect beyond the school, in this case to be prepared to share resources electronically.

In the FarNet project, along with being on a list with others who taught a subject, teachers were asked to post resources for others to access. This deceptively simple request of the teachers required a major shift in terms of deprivatising practice. Elmore (2000) points out that the traditional model of schooling does not readily allow for communication between the individual classrooms and the wider context within which teachers work. There is an emphasis on professional autonomy and the right of teachers to make detailed decisions about how and when the curriculum will be delivered and the methods by which their students will learn. Under this model, known as loose-coupling (Weick, 1976), teaching is seen as requiring a 'high degree of individual judgement'. This right of individual judgement, or professional autonomy, is closely guarded in many instances. The result is that teachers frequently work in 'isolated classrooms, under highly uncertain conditions' (Elmore, 2000, p. 6).

This model of loose-coupling has serious implications for the implementation and sustainability of learning communities within schools. Elmore (2000) suggests that it explains why most innovation in schools occurs in the structures around the school rather than within the classroom and why, where innovation does occur, it tends to be in isolated pockets and as the result of volunteerism. Volunteerism leads to innovations that are in tune with the personal values and dispositions of individual teachers (Elmore, 2000) rather than being connected to any collective goal or purpose such as that of the FarNet community as a whole. This concept of volunteerism can be seen among the curriculum leaders in FarNet who, without a strong guiding purpose and shared understanding of their role or of broader outcomes, focused on their own ideas and visions. One of the curriculum leaders felt his site had been successful because he 'had made the resource … [he] had his vision and liked what [he] was doing'. In reality, he would have created the resources regardless of FarNet; FarNet was just a convenient vehicle.

The issue for FarNet, and other such learning communities, is that teachers are often unwilling to share resources. As one curriculum leader stated 'one of the biggest things you battle with is a [lack of] willingness to share'. Only six of the nineteen curriculum pages on FarNet had resources supplied by other than the curriculum leader. Similarly, almost no one other than the FarNet managers or curriculum leaders posted to the lists. This reluctance can be attributed to at least two concerns. The first is that teachers feel protective of what they create because of the work involved and could 'feel ripped off if someone just comes in and whips that from under their nose'. This feeling is exacerbated between schools, where teachers may feel they are helping out another school with no subsequent benefit to them or their students. As one interviewee explained 'Teachers were saying, "This is the resource I made. I am not going to share it with X school. What have they done?" '

The second is a concern by some teachers that their work might not be good enough. Laying one's work open to scrutiny by colleagues requires a fairly high degree of self-efficacy and an ability to take risks and to accept criticism. One curriculum leader reported that when asked for resources to post, teachers responded, 'Oh, it is not really good enough and I don't want people to think that it is really bad'. Another mentioned the fact that 'teachers want their resources to be perfect'.

A willingness by teachers to share and to reveal elements of their practice requires a climate conducive to the operation of a professional community. Teachers are more willing to deprivatise their practice when they know the environment is safe, supportive and constructive (Grossman *et al.,* 2001). There is no simple checklist but central notions in building the 'set of obligations, opportunities and resources for teacher learning' (Little, 2000, p. 257) include system thinking or collective responsibility; forms of ongoing collegial interaction and environmental conditions like a supportive principal and social trust (Toole, 2002, cited in Toole and Louis, 2002).

Within one of the schools in FarNet, there was a strong focus on teaching and learning led by the principal, who was passionate about pedagogical change and meeting the needs of all students. As a result, a professional learning community was beginning to form within the school. Staff had professional development related to the integration of ICT and pedagogical change. Teachers were also encouraged and helped to create resources and place them on the school intranet for students and colleagues and a part of their appraisal involved the creation of ICT resources. They were willing to do so because they 'felt safe in our little community first'. The result was a willingness to put the same resources onto FarNet for a wider audience.

ICT horse
Our argument here is not that teachers lacked skill in the use of ICT but, rather, that they may have lacked specific skills or knowledge as well as held certain dispositions about online learning.

Towards the close of the FarNet project, it was clear that teachers were prepared to use the Internet to find resources. In 2003, 97 per cent had accessed the Internet in relation to professional work while a slightly lesser percentage reported accessing the FarNet site (81 per cent). These results, along with those from the ICTPD data suggest that the majority of teachers are relatively comfortable with accessing material online to use in their professional work. They were also moderately confident with email. In terms of the use of email, 41 per cent used email regularly and 37 per cent described themselves as confident users.

The inference, however, is that they were not as comfortable with the Internet or email lists as a tool for collaboration, as an online learning tool. Unfortunately, the baseline ICTPD survey (Ham, 2001) did not ask teachers about the level of use of computers for professional communication and collaboration, although it did ask about preferred PD activities in relation to ICT. About a third of participants were ambivalent about the use of listservs, with no feelings either way. Small percentages had strong views: 10 per cent said they would hate it and 13 per cent felt it had strong appeal. Overall, of the eleven types of PD suggested, listservs were ranked tenth, with only professional reading having a lower mean level of preference. Release time to discuss and translate new ideas and strategies into unit plans with the help of a mentor was clearly the most preferred PD, followed by on-the-spot support (both, it is interesting to note, likely to be best sourced from within the school community).

There was a different story in terms of overall integration of ICT into the classroom, with 32 per cent of respondents stating they had not yet blended ICT into their student learning activities and only six per cent stating that all, or almost all their units of work had an ICT component. Only 13 per cent of respondents reported they were able to create web pages using either HTML or an editing program while a third reported they were able to access information and follow links, and a quarter understood advanced search techniques.

General skill level in terms of use of ICT did not seem to be a limiting factor in terms of resource production and email communication. Rather, it was teachers' lack of classroom experience in the use of ICT generally, and web-based resources in particular and their likely limited knowledge of what might constitute an adequate electronic resource for the classroom that may have contributed to their reluctance to attempt to produce such for sharing. This may also partially explain the feeling that what they were able to produce may not be adequate. In terms of professional learning in ICT, there is some evidence that teachers were not favourably disposed to the use of the Internet as a tool for PD.

A 'need for' horse

From our data, there does not appear to have been widespread 'buy-in' to the notion of FarNet learning communities, in terms of such a community being perceived as something needed to solve existing problems. Indeed, in some ways FarNet appears to have been a solution looking for a problem. However, some did view what FarNet offered as a way of meeting their needs. In the FarNet project, the curriculum resources were, in part, envisaged as a supplement to teacher pedagogical content knowledge. In isolated areas with small numbers of senior students, teachers may 'teach' several subjects that are not part of their disciplinary background and training or, alternatively, may provide face-to-face support for students who take subjects by correspondence. For example, in one school, a senior teacher was responsible for seven different curriculum subjects at years twelve and thirteen and had accessed resources. Another example came from a teacher who had previously taught only primary school English and was now teaching senior English. FarNet has enabled this teacher to be mentored by an experienced teacher in a nearby school. Some smaller schools reported such 'little linkages' among themselves for support and the sharing of resources and ideas.

Maori teachers expressed the real need for them to be part of an electronically supported community: 'I think if we weren't sharing we would be lost, the Maori teachers if we don't share and get in touch with one another … Maori teachers used to feel isolated but with this, with FarNet, you don't'.

The bottom line in terms of building an electronic community such as FarNet was that teachers had to not only perceive the need but recognise that Internet resources, exchange and collaboration via FarNet were viable solutions to the perceived need. One curriculum leader interviewed saw FarNet as supporting those who wanted to teach in particular ways or move towards certain pedagogical approaches:

> It [FarNet] will go some way to being a catalyst to it [to change]. That is all it will ever do; it will make the job easier. But what has really got to come is the want to change to constructivism, [to be] experimental, where you have different layers inside your classroom – where you will have different activities inside your classroom, where it is differentiated between levels. If you want that to happen, FarNet is going to enable that to happen really big time … If you believe that you get better learning with a piece of chalk and a blackboard and you are going to give notes then don't get involved with Far Net.

Ensuring the Horses are Ready, Harnessed and   the Cart

The experience of FarNet suggests that there are a number of interrelated issues to address in order to maximise the likelihood of success of any online professional learning community. Considerable groundwork may be needed. As Fullan and Hargreaves (1996) note 'Some contrivance is necessary in the establishment of virtually all collaborative cultures. They don't happen by themselves' (1996, p. 58). First a clear need has to be identified for the electronic community. Further, there has to be a shared understanding of the value of the online community in meeting the need. FarNet, conceivably, could be a powerful tool to address several needs of teachers and their students in the Far North; however, the introduction and implementation did not ensure that ordinary teachers shared, let alone drove, the vision.

There are preconditions that enable or facilitate the development of professional communities and these include: openness to improvement (part of recognising the need for it); trust; mutual respect; availability of expertise; supportive leadership, and socialisation into the community (Kruse et al., 1995). The notion of collective learning and open consideration of practice should be developed at some level, as shown by one of the cases considered in FarNet, before expecting teachers to be willing to share aspects of their practice with a virtually unknown audience. In FarNet, the expertise available needed to be applied to the development of particular knowledge and skills required in the envisaged community – namely, those of constructing electronic resources, from the standpoint of experience of their use in the classroom and from knowledge of how to create them. Building a professional learning community is difficult to achieve within a school, let alone across schools, let alone virtually. Building on or strengthening an existing community is one way to approach this; supporting and guiding the building of communities within schools is another. Pursuing both of these avenues would enhance the impact of initiatives such as the vision of a FarNet learning community.

The authors were provided with an electronic copy of the raw data for the FarNet schools from the baseline ICTPD survey, which they analysed as part of their evaluation of the project.

References

Achinstein, B. (2002). Conflict amid community: The micropolitics of teacher collaboration. *Teachers College Record, 104* (3), 421–55.

Bakkenes, I., de Brabander, C. & Imants, J. (1999). Teacher isolation and communication network analysis in primary schools. *Education Administration Quarterly, 35* (2), 166–202.

Becker, H.J. (2001). *How Are Teachers Using Computers in Instruction?* Paper presented at the 2001 Meetings of the American Educational Research Association.

Cochran-Smith, M. & Lytle, S. (1999). Relationships of knowledge and practice: Teacher learning in communities. *Review of Research in Education, 24,* 249–305.

Cox, M., Abbott, C., Webb, *et al.* (2004). *ICT and attainment: A review of the research literature.* ICT in Schools Research and Evaluation series No. 17 (Becta).

Cox, M., Abbott, C., Webb, *et al.* (2004b). *ICT and pedagogy: A review of the research literature.* ICT in Schools Research and Evaluation series No. 18 (Becta).

Darling-Hammond, L. & Sykes, G. (eds) (2000). *Teaching as the learning profession: Handbook of policy and practice.* San Francisco, CA: Jossey-Bass.

Elmore, R.F. (2000). *Building a new structure for school leadership.* Washington DC: Albert Shanker Institute.

Fullan, M. & Hargreaves, A. (1996). *What's worth fighting for in your school?* New York: Teachers College Press.

Grossman, P., Wineburg, S. & Woolworth, S. (2001). Toward a theory of teacher community. *Teachers College Record, 103* (6), 942–1012.

Hargreaves, A. (1994). *Changing teachers, changing times: Teachers' work and culture in the postmodern age.* London: Cassell.

Kruse, S., Louis, K.S. & Bryk, A. (1995). An emerging framework for analyzing school-based professional community. In K.S. Louis, S. Kruse & associates (eds). *Professionalism and community: Perspectives on reforming urban schools.* Thousand Oaks, CA: Corwin Press, 23–44

Kulik, J.A. (1994). Meta-analytic studies of findings on computer based instruction. In E.L. Baker & J. H.F. O'Neil (eds), *Technology assessment in education and training.* Hillsdale, NJ: Lawrence Erlbaum.

Ham, V. (2001). *ICTPD Baseline Survey, 2001.* New Zealand: Ministry of Education

Little, J.W. (2000). Organizing schools for teacher learning. In L. Darling-Hammond & G. Sykes (eds). *Teaching as the learning profession: Handbook of policy and practice.* San Francisco, CA: Jossey-Bass, 233–262

Little, J.W. (1990). The persistence of privacy: Autonomy and initiative in teachers' professional relations. *Teachers College Record, 91* (4), 509–36,

Niemiec, R. & Walberg, H.J. (1987). Comparative effects of computer-assisted instruction: A synthesis of reviews. *Journal of Educational Computing Research, 3,* 19–37.

Parr, J.M. & Townsend, M.A.R. (2002). Environments, processes and mechanisms in peer learning. *International Journal of Educational Research, 37,* 403–23.

Schlager, M. & Fusco, J. (2003). Teacher professional development, technology and communities of practice: Are we putting the cart before the horse? *The Information Society, 19*, 203–20.

Selwyn, N. (2000). Creating a 'connected' community? Teachers' use of an electronic discussion group. *Teachers College Record, 102* (4), 750–78.

Shakeshaft, C. (1999). *Measurement issues with Instructional and Home Learning Technologies.* Paper presented at The Secretary's Conference on Educational Technology. Washington, USA.

Toole, J.C. & Louis, K.S. (2002). The role of professional learning communities in international education. In K. Leithwood & P. Hallinger (eds). *Second international handbook of educational leadership and administration*. Britain: Kluwer Academic Publishers, 245–79

Venezky, R. & Davis, C. (2002). *Quo vademus? The transformation of schooling in a networked world.* Retrieved 22 September 2003, from: <http://www.oecd.org/dataoecd/48/20/2073054.pdf>

Weick, K.E. (1976). Educational organizations as loosely-coupled systems. *Administrative Science Quarterly, 21* (1), 1–19.

Westheimer, J. (1998). *Among school teachers: Community, autonomy, and ideology in teachers' work.* New York: Teachers College Press.

The Use of Virtual Field Trips in the Classroom

9

Stephen Hovell

Introduction

Research has shown that virtual field trips (VFTs) provide valuable educational support (Butler, 2001; Graham *et al.,* 1997; Krupnick, 1998; Turturice, 2000) and suggests that they offer a valid use of technology in a real educational context (Bellan and Scheurman, 2001; Cisek, 2000; Hill, 1997; Riel, 1995). Research also indicates different activities that can make up a VFT, but little has been done on what makes their delivery effective in terms of what teachers do and how they are integrated into the classroom.

This chapter looks at how the Linking Education with Antarctic Research in New Zealand (LEARNZ) virtual field trips can be used to support the Science curriculum at Levels Two to Four. It begins by looking at current literature on VFTs. It shows how VFTs fit into the context of all online activities and discusses different uses and types of VFTs. It identifies characteristics of good VFTs and examines their effectiveness. It then goes on to discuss the LEARNZ VFTs and their various elements. This is followed by a discussion of what teachers did on the VFT and what they learned, drawing from my own research carried out in 2003. Lastly, there is a discussion on issues found when implementing VFTs and a list of recommendations is presented.

What is a Virtual Field Trip?

The terms 'virtual field trip' and 'electronic field trip' have been used interchangeably. Some teachers may take the term 'virtual' quite literally as meaning 'not real' – this creates a false impression of VFTs. This is why some prefer the term 'electronic' to signify delivery through an electronic medium. However, 'virtual' used in the sense of VFT simply means 'delivered via the Internet' (Wagner, 1998) or 'a digital alternative representation of reality' (Stainfield *et al.*, 2000), so the terms are synonymous. The professional teaching community is moving increasingly toward the term 'virtual field trip'.

The 2003 LEARNZ Teacher Manual acknowledges that VFTs exist in a multitude of forms. This becomes apparent when trying to define them. Some definitions, constructed by designers to align with the product they offer, may reflect a bias towards that specific type of VFT. Synthesising from the range of definitions, VFTs involve electronic travel via the Internet, beyond the classroom in either place or time, for the purpose of learning. Such a definition is supported by Cisek (2000) and Bellan & Scheurman (2001). Perhaps the best definition from a teacher's viewpoint is that offered by Nix (1999) as it both integrates and builds on the above. She identifies a VFT as 'an interrelated collection of images, supporting text and/or other media, delivered electronically via the World Wide Web, in a format that can be professionally presented to relate the essence of a visit to a time or place'. It reflects less bias than other definitions.

Herbert (1998) sees the Internet as a 'storehouse' for VFTs containing: the text; the interactive components; the multimedia components and the data, which is accumulated. It is the combination of these elements that makes the trip possible. There is a visual representation of the trip – photos or movie clips – using technologies such as digital camera, video camera or webcam. Chieng & Wong (in press) confirm that VFTs 'integrate textual, audio, graphic images and moving pictures into a single, computer-supported pedagogical product ready to be used in the classroom'. Beal & Mason (1999) emphasise the importance of establishing a 'sense of place' as an anchor and as a part of the student's own personal development. Electronic media is used in VFTs to place the student in a field-trip context outside their classroom, to embed them in an environment where, in learning about that environment, they are also learning about themselves.

Somerville (1999) subscribes to the view that VFTs need to have something that entices the learner and the teacher, something 'out of the ordinary'. This is recognised in exemplary VFTs that take students to offshore islands, explore remote national parks, follow dogsled races in Alaska or visit almost inaccessible Mayan civilisations. Thus Nix (1999) is correct in recognising that VFTs need to be fun and captivating. Further, Cowies (1997) sees VFTs as one of the most exciting applications of the Internet to teaching and learning. Enthusiasm for VFTs is evidenced in the literature. Graham et al. (1997) refer to the LEARNZ VFTs as 'an exciting and innovative educational experience' (p. 3). Graham even converted her classroom into an Antarctic scene. Turturice (2000) refers to the Internet (including VFTs) as an exciting vehicle that can be used to access resources not available otherwise. Krupnick (1998) conveys the excitement of VFTs in Dogsleds Online, a VFT that follows a dog-sled race in Alaska.

In 2000, a debate was held over the difference between the terms 'real' and 'virtual' in the context of natural history museums (NHM, 2000). Hawkey (2002) reports it soon became evident that such a dichotomy in terminology was false and the debate turned to the meaning of what is 'real'. Hawkey talks about the blurring of boundaries between the two. This argument, perhaps, reinforces why a number of organisations still prefer the name 'electronic' field trips over 'virtual' field trips and raises questions like 'how real is real?' and 'how virtual is virtual?'

In the evolution of VFTs, designers have tried to make the virtual experience as close as possible to a real experience. Nix (1999) affirms that VFTs present a worthwhile structure in which teachers and students can develop 'real science skills such as observation, inference, prediction, understanding and problem–solving.' The Passport to Knowledge: Live from … series relates what is going on in the classroom to what is happening in the real world, so that students can see that what they were studying was part of a larger picture. They wanted students to interact with experts, to work collaboratively with others and to have fun while learning (Riel, 1995). As with the LEARNZ VFTs and the Jason Project, they all have a teacher working in a real environment, access to online experts, and certainly the students are working collaboratively on real science and they have a lot of fun. LEARNZ aims at students doing 'REAL Science in REAL time with REAL people'. In a similar vein, the Jason Project uses the slogan, 'REAL Science, REAL Time, REAL Learning' while Passport to Knowledge uses 'REAL Science, REAL Scientists, REAL Locations, REAL Learning'.

Different Types and Uses of VFTs

There are different genres of VFTs, just as there are different genres in written language. One genre cannot be considered better or worse than another. The best VFT is one that meets the needs of a given situation. Some educators define these genres as different types, while others choose to differentiate VFTs according to use.

Far from replacing an actual field trip, VFTs have been used effectively to augment them. Bellan & Scheurman (2001) look on virtual and actual field trips 'as complementary components in a powerful instructional approach' to learning. VFTs can serve as preparatory pre-visit motivators. They can provide prior knowledge, trigger off questions, and generally focus attention on things students will see on the actual field trip. Orion & Hofstein (1994) believe that through pre-field trip activities, geographic, psychological and cognitive novelty factors that can distract students can be reduced, thus amplifying the value of the trip. This concept can be applied to VFTs as well (Worthington & Efferson, 1996). Stainfield *et al.* (2000) believe that through the use of VFTs prior to a real visit, students get the chance to develop some of the skills they will need, thus increasing the efficiency of the real trip. VFTs also serve as a post-visit activity for an actual field trip, to strengthen what the students discovered when they return to the classroom. It is an opportunity to synthesise, report on and even publish as their own VFT what the class has learned. VFTs can act as a replacement for the real experience, taking students to places they are unable to visit. (Beal and Mason, 1999; Bellan & Scheurman, 2001; Tuthill & Klemm, 2002). In Australia, the Minnamurra Rainforest Ecosystem VFT was designed with one purpose in mind: to provide students with the opportunity to compare another (virtual) environment with one they had already visited in real life as part of their scientific investigation (Placing & Fernandez, 2002).

Turturice (2000) discusses four types of VFTs: fact-finding missions (learning about countries, people, events and things); cultural awareness (using online museums to study cultural artefacts); concept application (applying theory learned in class to real world situations); and primary source exposure (a 'first-hand look at an actual event or issue'). Although these seem more pertinent to history, this classification can be applied to science in some contexts.

What Makes a Good VFT?

When the concept of actual field trips was first introduced, there was some reticence and suspicion among science teachers to adopt them. Stevenson (2001) contends that there have been parallels with the use of the Internet in the teaching of science. Placing & Fernandez (2002) report a similar resistance to participating in VFTs. Wood *et al.* (1997 cited in Placing & Fernandez, 2002) believe that this is because such teachers feel threatened by VFTs, seeing them as an attempt to bypass or even replace the traditional and valuable 'real' field trips that take place in so many schools. Stevenson (2001) argues that 'just as [real] field trips offer first-hand learning opportunities, so do VFTs'. Bellan & Scheurman (2001) recognise that technology cannot replicate 'the tactile, olfactory, visual and dialogical experience' (p. 160) of a real trip. VFTs are put forward as an alternative approach when a class visit to the actual site is not possible for any of a number of reasons, including distance, student welfare and environmental factors.

According to Mitchell & Wesolik (2002) and Tramline, VFTs provide access to a greater range of experiences and resources than do 'traditional' field trips. However not all teachers would accept this, preferring the hands-on actual field trip. Multimedia and current educational technologies have the potential to offer exciting new possibilities for learning and teaching, and lend themselves well to VFTs (Spicer & Stratford, 2001). Staley & MacKenzie (2000) are convinced that where face-to-face experiences cannot be achieved, for example, in an environmentally sensitive location such as a visit to look at endangered takahe or hazardous locations such as a volcanic lava flow, then technology can play a major role in creating a virtual substitute for the real experience. Indeed, such recreated experiences are almost impossible without the technology and, the authors argue, without the real or virtual experience, any learning would be abstract at best. They further suggest that very sophisticated learning experiences can be provided through using the new technologies.

It must be emphasised that VFTs 'are not inherently effective instructional tools' (Woerner, 1999). In other words, just because an activity is referred to as VFT does not mean it is educationally worthwhile. Just as physical field trips can be unprofitable so, too, can their virtual counterparts, falling into the category of entertainment (Qiu & Hubble, 2002) or 'a monumental waste of time' and acting as 'glorified travel brochures' (Bellan & Scheurman, 2001, p. 154). Their value lies in how they are used by the teacher. Some VFTs are definitely better than others. So what makes a good educational VFT?

A combination of the work of Nix (1999) and Yekovich et al. (1999) forms the basis of the following evaluation criteria. The term 'good' is used here to mean 'educationally effective'.

Good VFTs are goal-directed with links to the curriculum

Bellan & Scheurman (2001) report the increasing trend of educators to link curriculum objectives with Internet resources, suggesting VFT experiences are enhanced through such linkage. If such links are not present, the VFT can be regarded as nothing more than entertainment (Mitchell & Wesolik, 2002). It is now common to see sites such as the Virtual Geologic Field Trip to Griffith Park specifically developed to support the high school curriculum, listing relevant curriculum goals. It is also happening at the primary level – for example, Journeys to Wilderness Canyons was developed in the USA for Grade Four classes. Krupnick (1998) ties the excellent VFT, Dogsleds Online, to the American national curriculum objectives at a primary level. Locally, the Virtual Field Trip to Maungakiekie/One Tree Hill comprises five

VFTs that have been developed to meet requirements of Social Studies in the New Zealand curriculum.

Good VFTs are situated within an authentic context

Yekovich *et al.* (1999) observe that children come to the learning situation with some prior knowledge and this needs to be recognised. This previous knowledge serves as an aid in developing new knowledge. It is through experiencing such contexts that Worthington & Efferson (1996) believe students develop a sense of their own place in relation to the real world. They give an example of students visiting a hospital via a VFT and talking with a real doctor or nurse. This gives students some idea of how hospitals work and thus helps to refine their understanding of what hospitals and doctors do. It may also give insight into what is required if they want to become doctors or nurses themselves. Annenberg/CPB's Journey North is a VFT about global wildlife migration, where students are involved in tracking wildlife species such as the monarch butterfly across the North American continent. The LEARNZ Tongariro Volcanoes VFT is about something that is very real to New Zealand children. In scanning the list of VFTs offered by LEARNZ over the years, all themes have been authentic. This aspect also reinforces the importance of authenticity under the constructivist paradigm.

Good VFTs provide opportunities for the sharing of responsibility

This is about highlighting the social nature of learning. Rodrigues (2003) discussed the idea that much science teaching has involved the teacher 'holding the conversational floor' with the students in a passive role. This does not happen in good VFTs. The extent to which VFTs involve shared learning depends on the teacher's willingness to release control, again reinforcing the fact that ICTs work best in constructivist classrooms where learning is a shared experience. Responsibility in VFTs is shared among teacher, students, designers, facilitating teachers and online experts.

Good VFTs encourage multiple modes of expression

Modes of expression include oral, written, visual arts, drama, and so forth. This is another example of the designer-teacher-student partnership. The designer can make suggestions. It is up to the teacher to capitalise on these and extend such opportunities to benefit the students. The importance of such partnerships in the evaluation of software packages has been recognised by Squires & McDougal (1996).

Good VFTs use interactive multimedia technology

'Multimedia' refers to a range of technological media. 'Interactive' is used very loosely, but generally implies the user is in an active rather than a passive role. Interactivity can be visualised as lying along a continuum. At one end, it may involve simply clicking on a button, menu selection, or turning a page – what Plowman (1996) refers to as 'gratuitous interaction'. But interaction goes way beyond that. As Heppell (2001) points out, we interact with our toaster when making toast but that doesn't necessarily make it a worthwhile learning experience. The important element of interaction for Heppell is active participation. At the other end of the continuum, it may involve facilitating knowledge acquisition, skill development and understanding (Sims, 1997).

Pachler (2001) considers interactivity to be a defining feature of today's new technologies. In VFTs, interaction exists at different levels, an important aspect being the interaction between students and their teacher with some other person or people beyond the classroom. It may be a facilitating teacher as used by LEARNZ, an online expert, or even students in other classes. Sims (1997) suggests that the level of interactivity determines the depth of the learning experience.

Caulton (1998) considers interactivity in relation to virtual spaces as being based on clear educational objectives that allow the users to make choices, and use their initiative to explore the space. This exploration leads to developing understanding. In his geology courses, Butler (2001) uses interactive resources that engage students in constructing their own knowledge base. The word 'engage' implies working with materials, and using them purposefully. It is more than just exposure to them. Bilton-Ward (1997) acknowledges that interactiveness is one way of overcoming barriers that may arise through the virtual environment. This occurs as students become actively involved in learning. Accompanying the LEARNZ teacher to Auckland's volcanic cones, talking with a scientist about takahe, discussing ice formation in Antarctica with a glaciologist – all these encourage active participation. Activities do not need to be limited to the virtual environment. Other learning activities are encouraged in the classroom and local environment that parallel the work being undertaken in the VFT, and several teachers who do the VFTs follow up on this point.

Although there are many Web-based resources that are educationally inferior (Redfern & Naughton, 2002), this is not the case with quality VFTs such as those referred to in this chapter. For example, Meridian's Marshian Chronicles is 'a grassroots Internet field trip developed by a partnership of naturalists, educators, and a professional videographer/Internet enthusiast' about the estuaries of North Carolina. The developers found that, aside from the live interactive VFT being a different way of developing science concepts in the students, 'a new social context for communicating and relating is created' (Crissman, 2000). One of the reasons for this new social context was the high level of interactivity offered in the VFTs.

A good VFT encourages connectedness and interactivity in different ways: between the class and experts, between different classes involved in the trip, and between classes and a key teacher who is running the VFT (as with LEARNZ). Interaction may occur directly with material found within the site (Wilson, 1997) or away from the site, for example:

- by subscribing to bulletin boards as in the Iditarod Virtual Field Trip (Krupnik, 1998);
- through (asynchronous) electronic discussion forums where, for example, questions can be directed to students who may have actually visited the real site which is represented through the VFT (Jason Project student argonauts);
- through (synchronous) chatrooms or audio-conferencing especially set up for the VFT;

- through set classroom activities carried out away from the site and then reported back;
- through Internet links including hyperlinks, set up for student research;
- through email exchange between students, online teachers and experts.

Computer-mediated communication

This is a major feature of some genres of VFTs. Krupnick (1998) endorses the ability of the Internet to provide connections with other people as the best aspect of using online activities with students. For example, both synchronous and asynchronous CMC form an essential component in the LEARNZ VFTs. Examples of good practice involving synchronous communication are live audio- or video-conferencing. Such experiences allow students to interact with others and learn collaboratively. Students can interact directly with experts and facilitating teachers, or other people involved in the live aspect of the trip. Hill (1997), for example, records the success of interactive projects where one group of students works with another group at a distance. There is built-in opportunity for students in both groups to interact with researchers, archaeologists, explorers and historians. Asynchronous communication occurs through email and bulletin boards. The advantage with this type of CMC is that classes can access when they are able to, and are not limited in any way by time restraints, a problem that may arise with synchronous CMC.

Many VFTs compensate for their 'virtual' nature

This is accomplished by using a range of techniques to recreate a semblance of reality or a sense of place. A number of VFTs use *a live presence* (a facilitating teacher, student or other key person) whose task is to go on the real trip and portray it through the medium of the Internet to the students. LEARNZ uses a facilitating teacher. The facilitating teacher can also 'prepare resources, stage manage audioconferences, upload digital imagery and web pages and contribute to the listserv' (Sommerville, 1999). The students can contact the experts through this teacher. Anderson *et al.* (2001) endorse the importance of the 'e-teacher' claiming that they have a 'crucial' role to play in fostering online interaction. The Jason Series uses student 'argonauts' while Marshian Chronicles uses 'student hosts' to accompany a research team or naturalist to the real site of the VFT. Other techniques include the use of live video, audioconferencing in real time, and perhaps some sort of 'mascot' or 'ambassador' programme.

Good VFTs will offer additional support

For example, Woerner (1999) believes effective VFTs should provide concrete activities in preparation for the trip, and pertinent off-line and follow-up activities. Relevant background materials should be included. There is usually a separate teachers' guide either online or in hard-copy format. She further suggests that students should be able to move around the site at their own pace, and use those aspects of the site relevant to their needs. They should also have opportunity to discuss and compare the findings with other students and online experts. There should also be hyperlinks to related sites. Supplemental print material may also be available.

Guidance should be given in the form of technical help files. If the VFT uses videoclips, how can teachers download these? If there are any problems, how can they be resolved? Contact details should be readily available, including email addresses, phone or fax numbers. An online helpboard may be used where teachers and students can ask questions about problems they might be having.

The VFT website should reflect the elements of good website design

They should include most of the features of educationally effective websites, for example, as set out by Hovell *et al.* (2001). VFT developers should pay careful attention to content and evaluation, source/authorship and currency, design and structure, ease of navigation, performance of the site and judicious use of multimedia.

Good VFTs should be based on a sound pedagogy

Although listed last, it does not mean this is the least important factor. It is one of the most important. Developers of good VFTs should have a strongly developed awareness of current educational theory and this should be apparent at all stages in the design.

The decision over what constitutes a good VFT is not a simple matter of counting how many of the above criteria are included. Also, different factors may have different weightings. What is good for one class may not be considered good by another class. A good or effective VFT is one that meets the needs of the students, the teacher and the curriculum at that time.

Overview of the LEARNZ VFTs

In this section the LEARNZ field trips are used as an example to illustrate how VFTs work in practice. LEARNZ is 'an online education programme offering students learning experiences outside the classroom' (Somerville, 2003a) and traces its beginnings back to 1995. The first LEARNZ teacher, Pete Somerville, went to Antarctica under a Royal Society teaching fellowship, as part of the International Centre for Antarctic Information and Research (ICAIR) team. Having seen the US-based Live from Antarctica project, Somerville had a vision to see something similar working for New Zealand students. While maximising the use of the technology, he realised that whatever emerged needed to be pedagogically sound. His purpose was 'to investigate the relevance and possible application of an emerging technology called the Internet for New Zealand schools' (P. Somerville, personal communication). In his own words, Somerville stated, 'My belief was that the fascinating reality of Antarctic science could become an effective part of New Zealand classroom lessons, if driven by teachers ... The rest is history.'

Somerville's concern over pedagogy is laudable and, although he came to this conclusion eight years ago, it parallels the focus of the Ministry's 2002–2004 ICT strategy. Somerville acknowledges that creating stimulating learning opportunities is important, but this must be integrated into 'teacher-owned' classroom programmes (Somerville, 2003b). This is reflected in LEARNZ's first objective in the 2003 teacher manual, which is 'to develop and deliver quality net-safe e-learning resources focusing on the New Zealand Curriculum Framework' (p. 8).

LEARNZ is a Ministry of Education-funded Learning Experiences Outside the Classroom (LEOTC) project developed by Heurisko Ltd, with field-trip support from Solid Energy NZ Ltd and from the Department of Conservation/Te Papa Atawhai. The Royal Society of New Zealand has also been involved in previous years.

LEARNZ defines a VFT as 'a field trip that your class participates in by using classroom tools such as a computer and a telephone' (Sommerville, 2001, in Starters & Strategies, 2001, p. 39). The field trips take students outside their classroom and are consciously "designed to make the real world a part of normal every day teaching and learning" (Somerville, 2003a). The VFTs include: 'a comprehensive teacher manual, [a] special website area for teachers, clear curriculum links, aims and objectives written for [the teacher] … professional development for teachers integrating ICT into classroom practice, [and] an exciting and innovative way to learn that will motivate … students' (Starters & Strategies, 2001, p. 39).

LEARNZ VFTs are designed by practitioners and led by experienced teachers. They comprise a range of different components that can be '(re)assembled' by the classroom teacher to meet individual class needs. They offer live interaction with experts and give children a chance to meet people in the real world. The VFTs are best described through the whole LEARNZ experience, an example of which is available at the Rangitoto VFT site. This 'mini-site' reveals the format of a typical VFT which comes in two parts: preparation, and field trips.

Each year, at least one new LEARNZ teacher is appointed. This teacher acts as an interface between the class and their teacher, and the scientists or online experts. They are real people that the children can identify with (Graham *et al.,* 1997). Jonassen *et al.* (1999) state that learning takes place through thinking, and thinking develops through activity. The LEARNZ VFTs are about activity. As part of their role, the LEARNZ teachers try to create an environment that promotes active learning, engenders community and facilitates the learning process.

LEARNZ recognises the importance of professional development, so each year offers sessions for teachers to learn more about the VFTs. These are usually run by previous LEARNZ teachers who convey the message of LEARNZ very competently. They also serve to motivate teachers so that they are ready to take their class on a VFT with some relevant prior knowledge of the programme. As well as providing this pre-field trip training, teachers are also receiving 'on-the-job' training as they work alongside their students. LEARNZ recognises its role and responsibility through the following statement: 'LEARNZ has become increasingly recognised as a professional development resource for teachers. LEARNZ has built over the years a programme that supports every teacher regardless of their confidence with technology. Our programmes see teachers gaining confidence and becoming inventive and creative in integrating technologies into their teaching programmes' (Somerville, 2003b).

Schrage (1997) is of the opinion that the strength of new technologies lies not in their ability to deliver information but in their ability to foster relationships. Thus, he prefers to think of the current upsurge in the use of technologies as a 'relationship revolution' and not an 'information revolution'. This is where good VFTs such as LEARNZ come into their own. They provide a wide range of opportunities that strengthen relationships and, indeed, rely on relationships for much of their success.

All of the VFTs offered by LEARNZ since its inception are presented as integrated units and this is one of their great strengths. In terms of the curriculum, LEARNZ VFTs can be approached through five different threads:

- essential learning areas such as science or English;
- essential skills such as problem-solving and communication skills;
- environmental education – interdependence, sustainability, biodiversity, and personal and social responsibility;
- developing scientific skills and attitudes – focusing and planning, processing and interpreting;
- frameworks for learning which include concepts such as multiple intelligences, blooms taxonomy, de Bono's six hats, inquiry learning or Art Costa's characteristics of intelligent behaviour.

Components of a LEARNZ VFT

The best way to show what is involved in a typical VFT is to look at a graphic overview showing the components. The following is taken from the 2003 teacher guide.

Figure 1: Components of a LEARNZ VFT.

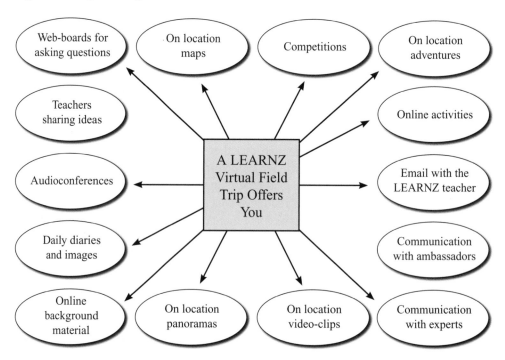

The VFTs are modular. Teachers can choose the components they want to use. LEARNZ is a microcosm of what can be done in ICT. It is about access to information and communication, and not about imposing a format. Schools are bound by the curriculum but the LEARNZ model allows some degree of self-determination as suggested by Lemke (1998).

One way of compensating for the virtual experience is through the *ambassador programme*. Classes are invited to send a small, soft toy to travel with the LEARNZ teacher on her journey. Each ambassador has its own webpage that shows photos of the ambassador on the VFT. The ambassador actually experiences the trip. One LEARNZ teacher remarked, 'We can't all go but we can send a little part of us.' Children identify with the VFT in this very real way. The class can email their ambassador and get a reply. The excitement on the faces of the students as they see their ambassador caught in a mist-net with Blue Ducks more than offsets the $25 it costs to send him on a trip. Teachers who use the ambassador programme affirm its success.

One of the most important elements is the *audioconference*. These run from twenty to twenty-five minutes and children can be involved either as a 'speaking school' or a 'listening school'. Audioconferences can be accessed using audiostreaming (listening via the website through the computer) or using a speakerphone. However, if a class is asking questions, a 'Polycom soundstation' is preferable, and the teacher manual gives extensive guidance on how to do this. Speaking schools must be well prepared. Children need to research their questions, and practise so that when it is finally their turn, they are able to speak clearly and confidently. Audioconferences allow students to ask questions of experts and also strengthen the relationships between the class, LEARNZ teacher and online experts. They also show students a valuable use of new technologies in an authentic context.

The *daily diary* is another essential element of the field trips. The LEARNZ teacher documents important events, each entry being accompanied by a photo to add the visual element. It is written in a friendly and interesting manner, which encourages the children to read. It also contains reminders for the class about competitions and other matters of interest. It is important that students follow the diary closely to keep abreast with what is happening. This is the thread that holds the whole trip together and keeps it moving forward.

LEARNZ uses three different sorts of *imagery* – the photo gallery, digital videoclips and panoramas – all of which can be accessed through the menu on the left-hand frame of each page of the VFT site. The photo gallery is searchable, with room to put in keywords from a scrollable list. All photos and panoramas first appear as thumbnails but can be enlarged with a single mouse click. Rural schools may find access slow due to slower download speeds but it is worth persevering to add this extra element of reality to the experience.

Schrum & Hong (2002) acknowledge the importance of having a place in online environments where students can ask each other or the (online) teacher for help. LEARNZ achieves this through its *web board*. Not only is it a place where questions are asked and answered, but it also reinforces virtual (online) relationships between all parties involved. Asking questions is an important part of the construction of knowledge.

One advantage of the LEARNZ *'experts'* is that they participate directly within the community of learners. The experts are more than the answerer of questions from students. They are partners in the learning process. Scardamalia & Bereiter (1998) advocate this type of involvement for greater effectiveness. Experts also help to show students that science is a part of everyday life and that there are opportunities for employment in the discipline of science.

Research into the LEARNZ VFTs

A study was carried out in 2003 to investigate the effectiveness of the LEARNZ VFTs in meeting the living world requirements of the New Zealand science curriculum at Levels Two to Four. The focus of this research was on how the VFTs supported the curriculum. Teachers in this study reported using the VFTs to meet achievement objectives (AOs) in the following areas and strands: science (living world, planet earth and beyond); social studies (time, continuity and change); technology (ICT – technological area); English (oral – listening and speaking, written – reading and writing, visual – presenting); and essential skills (communication, information, problem-solving, self-management and competitive, social and cooperative, and work and study skills). The majority of respondents used some form of integrated approach to the VFT. One ex-LEARNZ teacher suggested a key is that teachers do not look on the VFT as 'an extra' but as something that can be readily integrated into the classroom programme. Torrisi-Steele (2000) believes that the integration of ICT into the curriculum should be supported with other instructional tools and strategies.

All teachers reported on the VFT as a *learning experience*. All said that their students gained a lot in content knowledge: for example, how to identify takahe; why they are endangered; the location of the takahe recovery programme and how it works; and the importance of a bird's habitat to its survival. One of the strengths of the VFTs is the way teachers have used ICT as a part of the learning experience, and not relied totally on the online aspect of the VFT. For example, one teacher followed up the Aoraki VFT with a real field trip to Mount Cook/Aoraki. A second teacher has blue ducks not far from the school and she intended taking the class on a real visit as a follow-up to the VFT, showing how VFTs can work in tandem with actual field trips.

The participants were asked to rank the LEARNZ VFT as a learning experience for their class. Overall, teachers were highly supportive, the average ranking for the VFT experience being 9.2 out of 10. This means that the teachers who took part in the research felt that LEARNZ provided a very positive learning experience for their students.

Teachers were also asked to rank each of the specific learning outcomes (SLOs) for takahe and blue duck VFTs, based on the extent to which they felt they had been achieved. The average score was 9.3 out of 10, indicating how useful these teachers found the VFTs in meeting SLOs. The results show that when teachers completed the VFT, high outcomes were possible. One teacher put it like this: 'If teachers follow the VFT closely, they can't help but meet those objectives.'

Real science

LEARNZ uses the catch phrase, 'Real Science, Real People, Real Time, Real Locations'. In an effort to ascertain whether their children did real science in the VFT, teachers were asked to rate each of the twelve aims of science as stated in *Science in the New Zealand Curriculum* against what they did in the VFT. Teachers used a 1 to 10 scale with the option of using N/A if they felt the aim did not apply. The scores ranged between 7.4 and 9.0 and reflect the opinion all teachers held that the VFTs helped them do real science as defined by these aims. The reliability of using such an assessment tool is open to debate, which again lay outside the parameters of the study.

When asked to what extent the LEARNZ VFTs met the requirement of the science curriculum, all teachers scored this very highly (8.8 out of 10). However, most teachers recognised that LEARNZ was not the total package: it is merely one of many options teachers have for delivering the science curriculum. One teacher remarked that: 'there is a lot of science that a VFT can not actually do. The VFT can provide the motivation to carry on and do more. A lot of experiments and investigations cannot be done on the VFT. You do them in your own classroom. Some things do not come within the realms of a VFT. This is just one tool that is available to teachers.'

Another teacher delivered a seven-week unit on 'endangered species' with the VFT forming a part of that unit. For this teacher, LEARNZ is not a 'stand-alone' programme but forms part of other units that she is doing.

Other benefits

Aside from the content knowledge, there are other benefits for the students. At least two teachers reported that the children realised learning was fun through undertaking the VFTs. One teacher noted that her students had a real passion for learning and this served as a motivator for the teacher as well. This teacher remarked, 'You can go through the motions in some subjects – doing their reading, doing their maths – but through LEARNZ that passion for learning was realised.' A second teacher said, 'You can do reading and maths, core subjects which you can make exciting, but LEARNZ is something different and really exciting for the children to do. It makes learning so much more vital, alive and interesting.'

The range of activities in LEARNZ was far beyond what any one teacher could plan and organise individually, so it gave depth to the students' learning experience. The children in one class strengthened their computer skills, did map work and were encouraged to be independent workers. They acknowledged the level of the programme in the way they 'raved about it in their evaluations. They rated it TOPS.'

Several teachers noted the authentic nature of the VFTs as one of the advantages for their students, who gained a greater awareness of the world beyond the classroom. It widened the children's horizons and they felt as if they were actually participating in the field trip. Students learned 'about things which were not easily accessible, for example, looking at videoclips of the helicopter flying down the fiords, and talking to DOC experts in an audioconference.' Another teacher recognised that her students now have a much greater awareness of conservation issues and have been able to apply this knowledge locally. One teacher saw the benefits of taking the VFT and using it as a step to other 'real' projects beyond the classroom.

A number of teachers reported an improvement in essential skills by their students, particularly in the areas of problem-solving, social and cooperative, information, communication, work and study, and self-management skills.

As well as achieving outcomes in science, students gained a lot in the area of ICT-related skills. Participants reported their students as having learned skills in computer management, website navigation, Internet and email, using ICT in a real context to support other curriculum areas, specific software-related problem-solving, and skills with ICTs other than computers (for example, polycoms). Some of these benefits were not actually programmed into the VFT but happened as a result of the students using ICT in a real situation.

It is important that both teacher and students develop some technical skills: for example, basic keyboarding skills, how to store their work, how to manage files and print. Means & Olson (1995) refer to acquiring these basic skills as 'overhead' but, once learned, this investment can be applied across the whole curriculum. At the same time, it is acknowledged that although ICT skills are essential, 'doing the VFT is more than just using the computer as a tool – it is like a realm of its own.' It is not about learning the skills in a vacuum, but learning them through participating in an authentic rather than an artificial setting.

In discussing the concept of the 'intelligent school', Kate Myers, (from the Leadership for Learning Network at Cambridge University) reports the need to recognise the 'emotional element' involved in any new learning. She believes that learning involves the emotions before, during and after the learning experience (Morris & Hammonds, 2003). This emotional element was certainly commented upon by teachers in this study. When asked what sort of attitudes their students had towards the VFTs, teachers used words like positive, determined, enthusiastic, inspired and self-motivated. Most teachers reported VFTs eliciting a high level of interest among their classes. The enthusiasm may have derived from the VFT, or it may have been caught from the teacher, or a combination of the two. As one teacher commented, 'it's a different and exciting way to learn and the children love it.' When she asked her children for their opinion about VFTs, they replied with, 'awesome, really cool. I wish school could be like this all the time.'

Schrage's (1997) view of the new technologies fostering the 'relationship revolution' has been noted. It was pointed out that the LEARNZ VFTs provide a wide range of opportunities in the area of relationships. Participants shared this view in the way they spoke about the different relationships that existed while the VFTs were running. This is the 'real people' focus of LEARNZ in action. Special efforts are made by LEARNZ to develop and strengthen relationships. Although computers are sometimes thought of as being cold and inanimate, they do play a special role in developing relationships and one of the areas is through the LEARNZ teacher. This person needs to be 'positive, approachable and human' so that they are able to relate well with students, teachers and any online experts that appear in the programme. The children get to 'know' this person who becomes a part of their life. This is confirmed in what teachers described.

As they became increasingly involved in the programme, children develop a special relationship with the LEARNZ teacher. As one teacher put it, 'The kids think they know Audrie [the 2003 LEARNZ teacher] quite well, especially those who have been in the group more than once.'

The Role of the Teacher

The role of the teacher in implementing VFTs is crucial. The majority of participants cited the primary role of the teacher as that of facilitator. There are many different facets to facilitation including: motivating, encouraging, explaining what VFTs are all about, modelling good questioning, choosing the components the class will follow and generally how the VFT is delivered. It should also include providing opportunities for problem-solving, inquiry learning and fostering independence. Several teachers saw themselves as co-learners, acknowledging that the excitement of the teacher's learning carries across to the students. Being a co-learner also models a willingness to learn and a sharing between teacher and students.

In discussing the management of multimedia environments, Hill (2000) draws attention to the importance of 'pre-scaffolding', including what the teacher does before the online experience occurs. This preparation can be divided into two: 'teacher preparation' and 'student preparation'. The teacher must know how the VFT model works and be familiar with the website. One teacher concluded she had not prepared sufficiently: 'I learned at my own cost that I should have researched the website more carefully myself. As we went through the VFT I found out that I was learning alongside the children and I felt I should have been ahead of them.' Site familiarisation included understanding the background material so that it could be explained to students when they asked for assistance. It is important for the teacher to see the big picture and be aware of the different contexts that lead up to the VFT. Another teacher decided that on the next VFT she would ensure her children were better equipped by knowing more about the background of the animal being studied and, if sending an ambassador, understanding where it is going, what it would be doing, and what was their responsibility in supporting it. Students must also be aware of the learning outcomes so that they know what is expected of them.

Several teachers referred to their role in scaffolding. In today's classroom, scaffolding refers to the structure used by teachers to support their students' learning, to help them understand the material and reach their goal. It is not a fixture. It is taken away when the structure is no longer needed in relation to a specific learning task. Russell (1998) notes that teacher scaffolding has special significance in multimedia experiences, as it helps to improve learning outcomes. Scaffolding was identified by one teacher as important in developing prior knowledge, because it gives students something upon which they can build. McKenzie (1999) views scaffolding as: providing clear direction; clarifying purpose; keeping students on task; assessing to clarify expectations; pointing students to worthy sources; reducing uncertainty, surprise and disappointment; and delivering efficiently.

During the VFTs, most teachers used some sort of grouping to support collaborative learning. In several cases, group size was dependent on the number of computers available. Groups of three were often used: a technician who used the keyboard, a reader and a recorder to take notes. The positions were rotated so they all got a turn. While working in groups, teacher noticed how their children 'were feeding off each other. They were supporting each other and a lot of learning was taking place.'

Issues with LEARNZ VFTs

Each teacher in the research had her or his own individual challenges. These covered the technology, audioconferences, coming to terms with what VFTs are all about, the level of the teacher organisational skills and what to do with children who struggle with reading.

One teacher identified an overload problem, trying to balance the VFT with everything else that was happening in the school, but knowing the benefits for her students kept her going. Workload becomes an issue if a teacher tries to cover too many activities. Keeping one step ahead of the children can be difficult. Working out which components to use can present problems, especially for teachers new to VFTs. Another teacher found it quite challenging to relate the VFT to a younger age group and simplify the language (they are generally aimed at Levels Two to Four). Experienced users have reported withdrawal symptoms when they decide to miss out doing a VFT with their class.

Four areas of anxiety were raised: those of curriculum pressures from taking on too much; organisational issues such as keeping the new entrants quiet while older children were audioconferencing; rushing round to organise an audioconference; and ensuring everyone had filled in an acceptable-use policy form.

The majority of schools surveyed had accessibility issues, whether it be accessing from the classroom, a central computer room or a combination. The ratio of computers to students varied markedly from one to thirty to almost one to one in some cases. It was compounded by not all computers working. In considering labs versus classrooms, each arrangement has its particular advantages and disadvantages. The greater level of participation via the lab needs to be offset against the freedom of access in the classroom. Some teachers preferred the lab and others preferred the classroom, but best results were achieved using a combination of the two approaches, when available. It must be acknowledged that LEARNZ can and does work well with one computer. Teachers in this category tend to be creative in the way they use their single computer. Ideas include: downloading the information and printing a hardcopy; students checking throughout the day for new diary entries; integrating the VFT into the everyday learning of the classroom; and allowing groups of two to three to work on activities, using 'stations', two of which were computers. 'Other stations included Visual Language looking at posters, Reading stories associated with the topic and things like that. And it worked well!'

It appears that most schools also have their own stories when it comes to technical difficulties. Only one of the participants reported no technical problems. Most of the difficulties lay not in the LEARNZ site itself but in the hardware or software used to access LEARNZ activities. Three common areas of difficulty were identified: accessing VFT components such as videoclips; connection problems including network issues within the school and shortcomings with hardware. In reference to the 1996 LEARNZ expedition to Antarctica, Graham *et al.* (1997) observed that, 'It's not what equipment you have in the classroom, but what you do with it that makes the difference' (p. 3). Quite a number of schools still have old equipment with consequent reliability concerns. Slow access speed was an issue in some rural schools. But again, teachers were often creative in the ways they resolved these issues. In spite of the age of the computers, the fallibility of related hardware and networking problems, participation was not prevented. One teacher responded philosophically with 'You have to expect these things. You've got to cope with them.' This is emerging as a typical New Zealand teacher attitude when it comes to ICT.

Recommendations from Teachers

The participants have made a number of suggestions for teachers who might be undertaking a VFT for the first time.

Selection: Teachers should be selective about what they cover. Almost half of those surveyed raised this issue. The new teacher should choose one part to do first – perhaps the background reading on a specific topic, then bring in listening to audioconferences, sending an ambassador and emailing them, or following the diary. Another very experienced LEARNZ teacher suggested taking several elements of the programme and developing these. She added that teachers could reflect and refine their integration and try something different next time. It is important to take into account the needs of the students. One workable option was adapting the activities to suit individual class needs. This approach was certainly taken by a number of junior class teachers.

Preparation: It is important that teachers are well prepared before setting off on a VFT. Start by reading the teacher manual carefully. Visit the website early and read all the background pages. Teachers should spend time exploring the website for themselves before introducing it to the students. This is also an opportunity for the teacher to decide on the areas of focus. Becoming familiar with the site also helps the teacher carry out the function of guide. The better prepared the teacher is, the greater the potential for quality learning by the class. The children also need to be prepared before the LEARNZ teacher embarks on the physical trip; it is important to become familiar with the background pages and site navigation to maximise the VFT experience.

Time management: Accept that VFTs can be quite time-consuming. They can occupy a significant part of the teacher's time while the trip is on. One teacher declared that 'the VFT may just take over your life, especially if you want to take advantage of all there is on offer.' But in acknowledging this time commitment, one teacher does add a positive note stating that 'the teacher resource has ways of covering all aspects of the curriculum so it's possible not to feel guilty about spending a week in Fiordland or Mount Cook.'

Participation: Sending an ambassador is an excellent idea and gets the children more involved. It creates a sense of belonging. It also gets the teacher involved and was responsible for hooking at least one teacher into the programme.

Equipment: When undertaking an audioconference, teachers must ensure they have good equipment. Without it, the potential for management problems increases. Also, it is suggested that teachers have something for the children to draw or record on, especially if they have younger students. Pictorial recording is an option for Year Ones and Twos.

Inspect: Some teachers feel reticence towards computers, including VFTs. They might see one member of their staff using a VFT but have trouble taking the next step of actually using it themselves. Perhaps, as one respondent suggested, this might be stretching them outside their comfort zone. So one idea is to try and visit another teacher who is doing the VFT. Be a spectator. This helps the prospective teacher internalise what VFTs are all about.

Conclusion

This chapter investigated the use of the LEARNZ series of VFTs in integrating ICT into classroom programmes. It considered their ability to meet the requirements of the New Zealand curriculum, especially in science. The focus was on teacher perception and teacher practice as evidenced through questionnaires and interviews.

Findings from the study show that the LEARNZ VFTs are a quality resource incorporating elements that are found in the best VFTs offered through the Internet. The audioconferences, ambassador programme and web boards feature highly. Transcripts from this study considered the features of VFTs to be pedagogically sound as well as relevant to the New Zealand curriculum. They use a real presence in the form of the LEARNZ teacher and make a special effort at developing relationships with the students.

All teachers were very positive about participating in the VFTs. They affirmed strongly, the VFTs' ability to involve students in real science in authentic contexts and to address the AOs and SLOs in terms of the New Zealand curriculum. The VFTs reflect the current initiatives of the Ministry's ICT strategies and address concerns raised by the Education Review Office over the teaching of science and the use of ICT in classrooms.

The findings from the study suggest that the LEARNZ VFTs can support the integration of ICT into the New Zealand curriculum in a meaningful and exciting manner and are available at www.learnz.org.nz/downloads/dissertation-stephen-hovell.pdf.

References

Anderson, T., Rourke, L., Garrison, D.R. & Archer. W. (2001). Assessing teaching presence in a computer conferencing context. *Journal of Asynchronous Learning Networks, 5* (2), 1–17.

Beal, C. & Mason, C. (1999). Virtual fieldtripping: no permission notes needed. Creating a middle school classroom without walls. *Meridian 2*(1). Retrieved 9 October 2003 from: <http://www.ncsu.edu/meridian/jan99/vfieldtrip/index.html>.

Bellan, J.M. & Scheurman, G. (2001). Actual and virtual reality: Making the Most of Field Trips. In Stevens, R.L. (2001). *Homespun: Teaching local history in Grades 6–12.* Portsmouth NH: Heinemann. Retrieved 11 August 2003 from: <http://resources.heinemann.com/shared/onlineresources/E00334/chapter14.pdf>.

Bilton-Ward, A.C. (1997) Virtual teaching: An educator's guide. Retrieved 9 October 2003 from: <http://cord.org/vtc/virtualtech.htm>.

Butler, J. (2001) The virtual geosciences professor's course resources: good practices Retrieved 8 October 2003 from: <http://www.uh.edu/~jbutler/anon/interactive.html>.

Caulton, T. (1998). *Hands-on exhibitions: Managing interactive museums and science centers.* London: Routledge.

Chieng, A. and Wong, M. (in press). *The use of VFT for biological studies in Hong Kong Schools.*

Cisek, B. (2000) Travelling on the net: Virtual voyages. Retrieved 8 June 2003 from: <http://www.zapme.com/net/teacherslounge/teachtips/travelnet.html>. (link no longer active).

Cowies, S.K., (1997) Using the Web for virtual tours and skill-centered field trips: Lesson guide: Electronic field trips and other travel around the Web. National Institute for Literacy. Retrieved 8 October 2003 from: <http://novel.nifl.gov/susanc/eftskls.htm>.

Crissman, C. (2000). Marshian chronicles: Looking at the Past to See the Future of EstuaryLIVE. *Meridian 2000, 3* (1) Retrieved 8 October 2003 from: <http://www.ncsu.edu/meridian/2000wint/estuary/estuary6.html>.

De Bono, E. (1985). Six thinking hats. Boston: Little, Brown, and Co.

Gardiner, H. (1983). Frames of the mind: The theory of multiple intelligences. New York: Basic Books.

Graham, S., Donaldson, P., & Sommerville, P. (1997). Putting the curriculum on ice. *Computers in New Zealand Schools, 9*(2), 3–8.

Hawkey, R. (2002). The lifelong learning game: Season ticket or free transfer? *Computers & Education,* 38, 5–20.

Heppell, S. (2001). *New Zealand Education Gazette 80*(12), 8.

Herbert, B.E. (1998) Internet components of VFTs. Retrieved 8 October 2002 from: <http://trex.tamu.edu/faculty/herbert/98Golden/slide9.htm>.

Hill, M. (1997). *History-social science schools of California online resources for education (SCORE): one stop shopping for California's Social Studies teachers.* Retrieved 14 August 2002 from: <www.ccss.org/journal.htm>.

Hovell, S.R., Nicholson, E.S. & Fletcher, S. (2001). Development and evaluation of websites: A case study. In K.W. Lai. (Ed.), *e–learning: teaching and professional development with the Internet.* Dunedin: University of Otago Press, 109–128.

Jonassen, D.H., Peck, K.L. & Wilson, B.G. (1999). *Learning with technology: A constructivist perspective.* New Jersey: Prentice-Hall.

Krupnick, K. (1998). Dog sleds online: Creating a virtual field trip. *Social Studies Review, 38*(1), 43–46.

Lemke, J.L. (1998). Metamedia literacy: Transforming meanings and media. In Reinking, D., McKenna, M., Labbo, L., & Kieffer, R.D. (1998). (eds), *Handbook of literacy and technology: Transformations in a post-typographic world.* Hillsdale, NJ: Erlbaum, 283–301.

McKenzie, J. (1999). *Scaffolding for success.* Retrieved 20 October 2003 from: <http://www.fno.org/dec99/scaffold.html>.

Means, B. and Olsen K. (1995). *Technology's role in educational reform: Findings from a national study of innovating schools.* September 1995. Retrieved 4 August from: <www.ed.gov//PDFDocs/techrole.pdf>.

Mitchell, S. and Wesolik, F.J. (2002). *Virtual field trips for early and middle childhood educators.* Paper presented to the 18th Annual Conference on Distance Teaching and Learning, Madison, Wisconsin. August 14–16, 2002. Retrieved 29 September 2003 from: <www.uwex.edu/disted/conference/proceedings/DL2002_W3.pdf>.

Morris, W. and Hammonds, B. (2003). Leading – learning for the 21st century. *E–zine No. 15,* August 2003. Retrieved 26 August 2003 from <www.leading-learning.co.nz/newsletter.html>.

Nix, R.K. (1999) *A critical evaluation of science–related virtual field trips available on the world wide web.* Retrieved 4 August 2002 from <http://www.dallas.net/~rnix/vft_text.html>.

Orion, N. & Hofstein, A. (1994). Factors that influence learning during a scientific field trip in a Natural Environment. *Journal of Research in Science Teaching, 31,* 1097–1119.

Placing, K. & Fernandez, A. (2002) *Virtual experiences and the NSW stage 6 science Syllabus.* Retrieved 20 October 2003 from: <http://science.uniserve.edu.au/school/tutes/virtexps/paper.pdf>.

Plowman, L. (1996). Designing interactive media for schools: a review based on contextual observation. *Information Design Journal, 8*(3), 258–66.

Qiu, W. & Hubble, T. (2002). The advantages and disadvantages of virtual field trips in geoscience education. *The China Papers.* 75–79. Retrieved 20 October 2003 from: <http://science.uniserve. edu.au/pubs/china/vol1/weili.pdf>.

Redfern, S. & Naughton, N. (2002). Collaborative virtual environments to support communication and Community in internet-based distance education. *Journal of Information Technology Education, 1*(3) 201–11.

Riel, M. (1995). *Live from Antarctica: electronic travel and student interactions with distant mentors.* Paper presented at AERA, San Francisco. April 1995.

Rodrigues, S. (2003). Conditioned pupil disposition, autonomy, and effective use of ICT in science classrooms. *The Educational Forum, 67*(3), 266–75.

Schrage, M. (1997) *The relationship revolution: Understanding the essence of the digital age.* Paper presented to the Merrill Lynch Forum March 1997. Retrieved 8 August 2003 from: <http://www. ml.com/woml/forum/pdfs/relation.pdf>.

Schrum, L. & Hong, S. (2002). Dimensions and strategies for online success: voices from experienced educators. *Journal of Asynchronous Learning Networks, 6*(1), 57–67.

Sims, R. (1997). *Interactivity: A forgotten art?* Retrieved 21 July 2003 from: <http://www.gsu.edu/ ~wwwitr/docs/intrreract/>.

Sommerville , P. (1999). Virtual field trips. *The Good Teacher Term 1 1999.* Retrieved 20 October 20003 from: <http://www.theschoolquarterly.com/info_lit_archive/online_ict_learning/99_ps_vft.htm>.

Sommerville, P. (2001). Island Odyssey teacher manual. Christchurch: Heurisko.

Somerville, P. (2003a). LEARNZ Homepage. Retrieved 19 August 2003 from: <http://www.learnz.org. nz/>.

Somerville, P. (2003b). About LEARNZ. Retrieved 19 August 2003 from: <http:www.learnz.org.nz/ about.php>.

Spicer, J.I. and Stratford, J. (2001). Student perceptions of a virtual fieldtrip to replace a real field trip, *Journal of Computer Assisted Learning, 17,* 345–54.

Squires, D. & McDougal, A. (1996). Software evaluation: A situated approach. *Journal of Computer Assisted Learning,* 12, 146–61.

Stainfield, J., Fisher, P., Ford, B. & Solem, M. (2000). International virtual field trips: A new direction? *Journal of Geography in Higher Education, 24*(2), 255–62.

Stevenson, S. (2001). Discover and create your own field trips. *Multimedia Schools September 2001, 8*(4), 40–45. Retrieved 23 September 2003 from: <http://www.infotoday.com/MMSchools/sep01/ stevenson.htm>.

Turturice, M. (2000). Planning a virtual field trip. *Teaching History with Technology 1*(1) Fall 2000. Retrieved 4 August 2002 from: <http://www.caryacademy.org/historytech/Vol1no1/printable/ virrtrip.pdf>.

Tuthill, G. & Klemm, E.B. (2002). Virtual field trips: alternatives to actual field trips. *International Journal of Instructional Media, 29*(4), 453–68.

Wagner, E. (1998). Creating a virtual university in a traditional environment. Retrieved 10 August 2003 from: <http://www.uni-hildesheim.de/ZFW/vc/veroeff/veroeff-bologna.htm>.

Willis, A. (1999). Content–rich commercial websites. *Social Education 63*(3), 157–9.

Wilson, E.K. (1997) A trip to historic Philadelphia on the web. *Social Education,* 61 (3) 170–172.

Woerner, E. (1999). *Virtual field trips in the earth science classroom.* ERIC ED446901.

Wood, J. *et al.* (1997) Designing a virtual field course? *Eurographics 97,* University of East Anglia. Retrieved from: <http://www.geog.le.ac.uk/jwo/research/conferences/Eurographics97/index.htm>. (Link no longer active).

Worthington, V. & Efferson, N. (1996). *Electronic field trips: Theoretical rationale.* Retrieved 13 July 2003 from: <http://commtechlab.msu.edu/sites/letsnet/noframes/bigideas/B1/b1theor.html>.

Yekovich, F.R., Walker, C.H. & Nagy-rado, A. (1999). *Building TRALES to literacy for young learners.* Laboratory for Student Success No.402. Retrieved 7 August 2003 from: <www.temple.edu/LSS/pdf/spotlights/400/spot402.pdf>.

Websites

Jason Project: <http://www.jason.org/>.

Journey North: A Global Study of Wildlife Migration:<http://www.learner.org/jnorth/>.

LEARNZ Home Page: <http://www.learnz.org.nz>.

LEARNZ Tongariro Volcanoes: <http://www.learnz.org.nz/trips/tongariro_volcanoes.php>.

Marshian Chronicles: <http://www.ncsu.edu/meridian/2000wint/estuary/estuary6.html>.

MayaQuest (VFT site no longer active, but details can be found at): <http://www.concord.org/newsletter/1997spring/mayaexpedition.html>.

Passport to Knowledge: <http://passporttoknowledge.com/main.html>.

Virtual Filed Trip to Griffith Park: <http://www.laep.org/target/technology/secondary/griffith/>.

Virtual Field Trip to Maungakiekie/One Tree Hill: <http://www.tki.org.nz/r/socialscience/curriculum/SSOL/onetreehill/index_e.php>.

Searching for Information from the Web

10

Keryn Pratt

Introduction

Recently, a series of studies have looked at the effects of an ICT hardware grant in the Otago region (Lai and Pratt, 2003; Lai *et al.*, 2001; 2002). As part of these studies, information was gathered on the availability, and students' use, of computers and the Internet in Otago primary and secondary schools. In 1997, secondary schools had an average computer to student ratio of 1:11.6 (range 1:7 to 1:23), with this improving to 1:6.9 (1:4–1:13) in 2000 and 1:6.3 (1:2.1–1:11) in 2001 (Lai *et al.*, 2001, 2002). In 2002 the ratio of computers to students in primary schools was 1:9.5, with 98 per cent of teachers and 90 per cent of ICT coordinators having at least one computer in their classroom (Lai and Pratt, 2003). On average, 63 per cent of student computers in secondary schools had access to the Internet in 2000, with this improving to 78.4 per cent in 2001 (Lai *et al.*, 2001, 2002). Nearly 99 per cent of senior and 96 per cent of junior students reported they had access to the Internet at school in 2000, with all but 0.3 per cent of students being able to access the Internet at school in 2001 (Lai *et al.*, 2001, 2002). The majority (86 per cent) of student computers at primary schools were connected to the Internet, with 12 per cent of Years Six to Eight students reporting they never used the Internet at school, 41 per cent that they hardly ever did so, 26 per cent that they did so once a week, 19 per cent using it most days and 2 per cent doing so every day (Lai and Pratt, 2003). In 2000 most students, 76 per cent of junior and 81 per cent of senior secondary students, had access to computers at home, with 60 per cent and 69 per cent having home access to the Internet (Lai *et al.*, 2001). In 2001 the majority of students (90.5 per cent) with a computer at home had access to the Internet, with over half of these students (53.4 per cent) using it at least daily (Lai *et al.*, 2002).

In these Otago schools, the use of ICT, and in particular the use of the Internet, increased dramatically. This may be in part due to an increase in teachers' beliefs in the value of the Internet. In 2000, 88 per cent of teachers, 90 per cent of ICT coordinators and 95 per cent of principals indicated that it was important that 'students have access to the World Wide Web (WWW) as a learning resource' (Lai *et al.*, 2001). For example, in 1997, 26 per cent of secondary teachers required students to use the Internet, with this increasing to 88 per cent in 2000 (Lai *et al.*, 2001). This pattern continued, with 29.7 per cent of teachers using the Internet to search for information with their students at least weekly, and 45.2 per cent expecting their students to do so independently (Lai *et al.*, 2001). Indeed, over half the secondary students (51.5 per cent) reported doing research using the Internet for school at least once per week, while one third of students reported that they used school computers for research on the Web during their free time in 2000 (Lai *et al.*, 2001). At the primary level, 58.6 per cent of teachers and ICT coordinators reported occasionally or sometimes using ICT with their students for searching for information and 34 per cent doing so mostly or all the time (Lai & Pratt, 2003). The majority of primary students (52.4 per cent) also reported using school computers to find information on the Internet at least once a week (Lai & Pratt, 2003). Students also used the Internet at home to find information for school, with 54.5 per cent of secondary and 33.8 per cent of primary students doing so weekly and 11.3 per cent of secondary and 30.3 per cent of primary students doing so daily (Lai *et al.*, 2001, 2002; Lai & Pratt, 2003). Similar uses are also documented in other studies. For example, Smerdon *et al.* (2000) found that the most frequently assigned use of computers by teachers was for word processing or creating spreadsheets and for Internet research and Becker (1999) found that the most common uses of ICT by United States students was word processing, CD-Roms and using the Internet for 'research'.

How do People Search for Information on the Web?

Until recently, very little has been known with regard to how people use search engines to find information on the Internet. In 2000 Jansen and Pooch published a 'review of all Web-searching studies that deal with studies of searching using Web search engines' (p. 237), reporting on three such studies. A number of studies exploring the use of search engines to find information other than those included in Jansen and Pooch's review have been published (for example, Bilal, 2000, 2001, 2002; Nachmias & Gilmad, 2002; Schacter *et al.*, 1998; Wallace & Kupperman, 1997). In 2002, Nachmias and Gilmad explored how fifty-four graduate students, who were regular users of the Internet, found information on the Internet. These students were given a review of online search methods and a list of commonly used search engines before their performance at three search tasks (finding a picture of the Mona Lisa, finding a complete text of Robinson Crusoe or David Copperfield and finding an apple pie recipe that was accompanied by a picture) was tracked and analysed. Students were able to complete the tasks in any order, and there was no time limit. Only 15 per cent of students successfully completed all three tasks, with 39 per cent completing one task, 42 per cent completing two tasks, and 6 per cent of students failing to successfully complete any of the three tasks. Students spent between one and fifty-six minutes on each task. From their analysis of the search behaviours of these students, Nachmias and Gilmad (2002) identified three general types of search strategy, as well as a number of subtypes, used by students in attempting these tasks (see Table 2). The students in this study used search engine strategies almost three times more than they did browsing strategies, and tended to use simpler strategies (for example, keyword searches) most frequently, and complex strategies (for example, Boolean searches) rarely. This is supported by the finding that the most common strategy was the use of a single keyword within a search engine.

Table 2: Participants' search strategies as identified by Nachmias & Gilmad (2002; p. 481).

Strategy	Description	Example
Search engine strategies		
Keyword search	Direct typing of the query subject	Typing the words 'Mona Lisa'
Wide search definition	Searching using a broad query	Searching for art and painting to find the Mona Lisa
Complex search	Cross-searching with more than one keyword	'Picture' 'Mona Lisa' 'Louvre'
Use of general knowledge	Using information that is not mentioned in the search task	Searching for the Mona Lisa mentioning Leonardo Da Vinci
Computer convention	Using a computer convention	File suffixes (e.g. gif, jpeg)
Boolean search	Using Boolean syntax	Louvre AND Mona Lisa
Browsing strategies		
Using a directory	Browsing through a directory or catalogue	*Yahoo*! directory of topics
Accessing a specific portal	Look for the subject of interest (requires preliminary knowledge)	www.artnews.com
Direct-access strategy		
Direct typing	Simply type a URL	www.monalisa.com

Children searching for information

Schacter *et al.* (1998) noted that 'children's information seeking and use of the Internet are virtually unexplored areas' (p. 840). They explored the search behaviour of thirty-two Fifth and Sixth grade students (equivalent to Year Six and Seven students in New Zealand) who were asked to complete two search tasks (finding the three types of crime that happened most in California and finding information regarding what should be done to reduce crime in California). Children viewed an instructional software tutorial on the use of Netscape, and then had twenty-five minutes to complete each task, with task order counterbalanced. The children generally did not perform well on the tasks, with only two children correctly identifying the three most common types of crimes in California. Almost all children (thirty of thirty-two) found at least some information related to ways to reduce crime; however, the information they found was generally incomplete in terms of fully addressing the issue. The majority of students (twenty-one of thirty-two) used only one search engine, with eleven using more than one (this difference did not reach statistical significance). Most children (twenty of thirty-two) used natural language in their queries, entering complete sentences rather than keywords. None of the children used complex search strategies such as Boolean language or phrase searching, and they did not appear to plan their search strategies. Schacter *et al.* (1998) divided children's search behaviours into three types:

analytic   entering queries
browsing   visiting/revisiting webpages (including use of forward, back and reload buttons)
scan-and-select   returning to results page from a search

They found that over 80 per cent of children's information-seeking behaviour was browsing, with scan-and-select behaviour being the next most common.

Wallace and Kupperman (1997) observed four pairs of Sixth grade (Year Seven) students who were undertaking Internet searches looking for the answer to a question they had formulated as part of a week-long activity in a work unit on ecology. The students they observed did not use feedback from the search engine to improve or refine their searches. They noted that these students tended to feel they were succeeding if they could reduce the number of results or hits that were returned, and they sought suitable answers to the questions they were posing, rather than understanding. The pairs of students in this study did not explore widely, never going further than five links away from the original search page, and spending, on average, less than one minute per page. Although the students had been part of a class lesson on Web searching, they only used the basic functions of the search engine in their actual search.

Like Wallace and Kupperman (1997) and Schacter *et al.* (1998), Large and Beheshti (2000) found that, even with training, students generally only used the basic features of search engines, with most not utilising techniques such as the use of quotation marks to search for an exact phrase. Their study of fifty-three 12-year-olds found that the students had difficulty determining what search terms to use, and frequently became frustrated at the number of results, and especially the number of inappropriate results. They noted that the students tended not to use online help that was available, instead relying on other students and their teacher.

Bilal (2000, 2001, 2002) has published a series of articles based on her investigation of Seventh Grade (Year Eight) science students use of Yahooligans!, a search engine designed for children that allows both keyword and subject category searching. Ninety students were invited to participate, with twenty-two eventually taking part once parental and student consent was obtained, and three used for piloting. However, the number of students who participated in each task, and whose results were available for each task, varied somewhat due to absences and technical failures. The students undertook three searching tasks, each with a different characteristic (see Table 3). Children undertook the tasks in the order they are presented on the table, on the Monday, Wednesday and Friday of one week. They were given no instructions about how to do the tasks, or in how to use the search engine, but were encouraged to ask questions as needed.

Table 3: Description of tasks used in Bilal's series of studies of seventh-grade science students' use of Yahooligans!

Task type	Paper	Task	Time limit	Results based on
Factual	Bilal (2000)	How long do alligators live in the wild and how long in captivity?	30 mins	14 students
Self-generated	Bilal (2002)	Students choose a topic of interest, then a specific topic or question	45 mins	15 students
Research	Bilal (2001)	How is ozone depletion affecting forests?	30 mins	13 students

In Bilal's (2000, 2001, 2002) studies, half of the children (50 per cent) succeeded in finding the correct answer to the factual question, and 69 per cent of the children were partially successful at the research task, with the remaining 31 per cent failing to find any information relevant to the task. The majority of children (73 per cent) were successful at finding the information they wanted in the self-generated task, with the remaining children either unable to find the information, or unable to decide on an aspect of the topic for which to look for information. Students tended to use keyword searching for their initial search, rather than looking at the subject hierarchies, with 64 per cent, 87 per cent and 69 per cent of students using keywords for the factual, self-generated and research tasks respectively. Students varied in the number of moves they subsequently made – for example, making between six and twenty-eight moves in the factual task. Children also repeatedly backtracked, looped (returned to websites they had previously been to, or repeated a search they had already done) and followed hyperlinks in their searches. They made a number of spelling mistakes when undertaking keyword searches, and some children also used natural language for their queries. None used Boolean operators.

Throughout these studies, children searched for information on a variety of tasks. Generally, they used simple one-word keyword searches, using few advanced strategies (such as Boolean commands or phrase searching). Some used natural language, some made spelling mistakes, and none appeared to have a planned search strategy. However, despite these limitations, these studies show that children can use search engines to find a range of information, although their searches are not always successful. The question of interest, then, is what factors play a role in determining whether searchers are able to locate the information for which they were looking? The literature has identified a number of factors that affect search performance, and these can be divided into two broad categories: first, the skills and knowledge that are necessary for successful searching; and second, factors associated with all aspects of the search, the searcher, the search engine and the search topic that may affect search performance.

In a recent study of children's searching behaviours using the Internet and a model of the Internet (see Lai & Pratt, 2004), the ability of Year Four (around eight to nine years old) and Year Eight (around twelve to thirteen years old) Otago and Southland children to find specific information regarding the kiwi, and the town of Kingston, New Zealand, was explored, with the findings in line with the literature. Year Four students were asked to find how the kiwi got its name, where to stay in Kingston, and the name of the ride for which Kingston was famous. Year Eight students were asked to find this information, as well as three types of kiwi and one place they might make their nest, and the cost for a family to go on the Kingston ride. Year Four students performed poorly, with fewer than one fifth of children being able to find the correct answers to the questions, while between 48 per cent and 87 per cent of Year 8 students identified the correct answers. The majority of students at both year levels used combinations of keywords in their searching, with very few using the search buttons within sites. Around half of the Year Four students read the summaries in the results pages to decide which sites to open, however fewer than one third of Year Eight students did so.

The Year Four children entered 88 queries, with 67 being unique, and the Year Eight children 132 (92 unique) while answering the questions. At Year Four the average number of words per query was 4.1, with this being 3.2 for Year Eight students. Over half the queries entered by Year Four students and nearly one third of those by Year Eight students used natural language, while just over 10 per cent of queries at each year level contained spelling or grammatical errors. The queries used by some children indicated that they failed to understand how the Internet worked. One student put in their potential answers, so that thinking that Kingston's famous ride might be a roller coaster they used this as their query. Other queries of this nature were entered – for example 'like a kiwifruit' (how kiwi got their name). Another student entered the keyword 'kiwi' and then followed this by entering the word 'name'. Upon looking at the results these students were surprised to see that the results were not limited to sites that had appeared in their first search and that also matched the criteria of their second search. From the comments of other children it was clear that they understood how the Internet worked – for example, a number initially entered 'Kingston', looked at the results, and then went back and added 'New Zealand', explaining that the computer needed to know which Kingston they wanted. Similarly, several students modified their initial kiwi search from 'kiwi' to some variation including both 'kiwi' and 'bird', explaining that the computer thought they meant the people.

Lazonder and colleagues (Lazonder, 2000; Lazonder et al., 2000) believed that searching for information using the Internet required two different types of skills: locating the correct site and locating the information within the site. They tested this by asking 25 fourth-graders (Year Five), who were classified as either novice or expert Internet users to complete three search tasks. As predicted, experts were faster at locating sites, completed more tasks successfully and needed less time and fewer actions to do so. There were, however, no differences between experts' and novices' performances at finding information on a site, once it had been located.

In their recent study of children's searching behaviour on the Internet, Lai and Pratt (2004) found that although Year Four students found sites with links to other sites containing the answer on 7 occasions, they never followed these links. In contrast, Year Eight students entered sites containing links on 28 occasions while answering the kiwi question, and on thirty-three occasions while answering the Kingston questions, and followed these links on 16 and 32 occasions, respectively. Only 11 Year Four and 27 Year Eight students found the answer to the kiwi name question, although 31 Year Four and 41 Year Eight students entered sites that contained this information. Similarly, only 12 of 18 Year Four students identified somewhere to stay in Kingston, and 38 of 43 Year Eight students identified the types of kiwi, while 31 of 40 correctly identified where they made their nests. However, the majority of Year Four students who found sites with the name of the famous ride correctly identified the information (11 of 12), as did all of their Year Eight counterparts. Most of the Year Eight students who found websites containing information relating to somewhere to stay in Kingston and the cost of the ride were able to correctly identify the information (30 of 31 and 32 of 36, respectively).

Factors that may affect Search Performance

A number of factors associated with the searcher, the search engine and the search topic have been identified as affecting either the success of a search or how the search is carried out. These include: the type of task; limitations of the search engine; and individual characteristics such as gender, study approach, cognitive style and perceptions of, and experience with, the Internet.

Type of task

Bilal's series of studies compared children's search performance across three different types of tasks: factual, research, and self-generated. As measured by their success rates, children found the research task more difficult than the factual task (Bilal, 2001), and the self-generated task easier than either of the other tasks (Bilal, 2002). Children also conducted more searches in the self-generated task than in the research task, and the most searches in the factual task. When searching for the self-generated task, children were more successful when they browsed than when they used keywords (Bilal, 2002). Bilal (2000, 2001) also found that children who were unsuccessful at the tasks used natural language, and did more looping, more moves, scrolled less, used fewer hyperlinks and entered fewer homepages than did the children who successfully completed the task.

Schacter *et al.* (1998) also explored whether the type of task affected the success rates of fifth- and sixth-grade (Year Six and Seven) children's searching. They looked at children's search performance on a 'well-defined' finding and an 'ill-defined' searching task (finding the three types of crime that happened most in California and finding information regarding what should be done to reduce crime in California, respectively). Children performed much better on the ill-defined task (30 out of 32 finding some information) than the well-defined task (16 out of 32 finding some information). Overall, children used browsing (visiting/revisiting webpages) more than either analytic (entering queries) or scan/select (returning to results page) behaviours, and used analytic searches more on the well-defined than the ill-defined tasks. However, children found more information, and more relevant information, when searching on the ill-defined task. This may be, in part at least, due to the wording of the question that instructed them to find three pieces of information for the well-defined task, meaning children reported only three pieces of information. Schacter *et al.* (1998) also looked at children's perceptions of how well they had completed the tasks, comparing these with the ratings of experts. They found that experts and children did not differ on how they rated the children's performance on the ill-defined task; however, children overestimated how well they had performed on the well-defined task. The authors also found that gender affected the children's search behaviour in these two tasks. Overall, boys undertook more information seeking behaviours than did girls; however, girls browsed more than boys. Also, girls used more analytic behaviour than boys in the ill-defined task, but not in the well-defined task.

Individual differences

In addition to gender differences in search behaviour, other individual differences have been found. Ford *et al.* (2001) explored the relationships between the search performance of sixty-nine masters' students and individual differences (cognitive styles, Web and search engine experience, perceptions of the Internet, study approaches, age and gender). Their initial results suggested that older students, males and those with low levels of cognitive complexity were more effective in their search performance, while Internet experience made no difference. Further exploration of the results showed that age and gender were significantly correlated, with gender being the factor that was related to search performance, rather than age. They also found a number of other relationships, with a poor search performance linked to specific cognitive styles, study approaches and perceptions of the Internet. Palmquist and Kim (2000) also explored the role of cognitive style on search behaviour, and reported that it had an effect only if the searcher was a novice. In order to reach this conclusion, however, they raised the significance level to 0.1, so their finding must be treated with caution.

Experience

In Bilal's (2000, 2001) studies of children's search performance of three different types of tasks, unsuccessful children had less Internet experience and less knowledge of the search engine than the successful children. Children's domain and topic knowledge and their reading ability were not, however, related to whether their search was successful or not. Lazonder and colleagues (Lazonder, 2000; Lazonder *et al.*, 2000) took a different approach: instead of looking at search performance on different tasks, they looked at search performances during different stages of the search, and the effect experience had on searchers' performance at each stage. Lazonder divided Web searching into two phases – locating the site and locating the information within the site – and maintains that each phase includes four activities: 'goal formation, strategy selection, strategy execution, and monitoring' (Lazonder, 2000, p. 327). Figure 1 shows this process pictorially.

Figure 1: Process model of information searching on the WWW (adapted from Lazonder, 2000, p. 327).

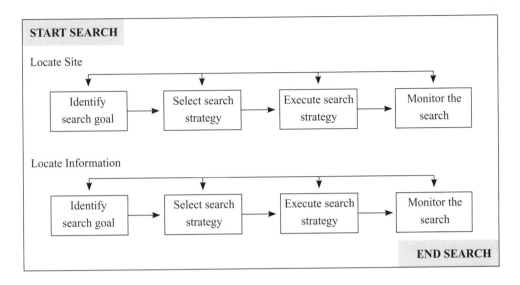

In the first study, Lazonder (2000) explored fourth-grade (Year Five) students' search performance in the first phase of this model, locating the site. Students were classified as novice (fewer than 10 hours of experience, not particularly proficient) or experts (over 50 hours of experience, fairly proficient) and were asked to complete three tasks of varying complexity. The complexity of the task was determined by the 'level of inferencing required to deduce the site's URL from the task description' (Lazonder, 2000, p. 330). In the low-complexity task the URL was given in the task, in the medium-complexity the URL could be inferred from the task, and in the high-complexity task the URL could not be inferred. Being given the tasks meant the first step of the process, identifying the goal, had already been completed, so performance was assessed on the remaining steps of strategy selection, strategy execution and monitoring.

Lazonder (2000) found that experience affected the initial strategy selection step of the process, with experts requiring less time to select strategies, and selecting better strategies, in the low- and medium-complexity tasks. There were no differences in strategy selection at task three, the most complex task, probably because the task didn't contain references to a site, and so experience at using the Internet would not be of benefit. At the third step, experts were faster at executing the strategy in task one, with no differences in tasks two or three. Further investigation showed that experts were faster than novices in the first task because they made fewer mistakes than novices. By the second and third tasks, however, novices made fewer errors, and spent less time exploring the search engine – probably because they had already explored the system and discovered some common errors as they completed the first task. There were few differences in the fourth step, monitoring, although experts were faster overall at task three. Novices' performance on task three was widely varied, and there were few, if any, systematic differences between novices and experts in the way they approached the search. Lazonder (2000) concluded that although the groups differed on their system knowledge and skills, which affected the strategy selection and execution steps, they did not differ in their information-searching skills, as shown in the monitoring step. He also noted that none of the participants used the search engine to its potential, or took full advantage of the information provided in results screens.

In a study further exploring the relationship between Internet experience and its effect on different stages of searching, Lazonder *et al.* (2000) compared the performance of experts and novices at locating sites and locating information on sites. Experts and novices were classified as they were in the previous study, and were asked to complete the same tasks. In line with Lazonder's (2000) conclusions, experts were faster at locating sites, completed more tasks successfully and needed less time and fewer actions to do so, than novices. There were, however, no differences between experts' and novices' performances at finding information on a site, once it had been located.

Search engines

In her studies of children's performance on a series of tasks using *Yahooligans!* Bilal (2000, 2001, 2002) identified a number of ways in which the search engine itself, and its design, affected children's search performances. As she noted: 'Lack of search instructions, search examples, and error recovery methods from both *Yahooligans!*' search and retrieval interfaces increased search repetition under inappropriate syntax. In addition, the limited instructions and guidance provided under *Yahooligans!*' online *Help* compounded children's retrieval problems'.

Bilal (2000, 2001, 2002) noted a number of others ways in which *Yahooligans!* affected children's performance, including the lack of a spell-check or correction function, and a confusing and cluttered screen. In addition, in the *Yahooligans!* home page the keyword search is presented above the directory options, despite the fact it is designed as a directory-based search engine. This, Bilal (2000, 2001, 2002) believes, may have encouraged children to use keywords rather than the directory options.

It is apparent, from these studies, that children's performance on Internet searching tasks is dependent on a number of factors, related to the child, the task to be performed and the search engine used. Children's knowledge and skill at researching and using ICT, as well as their general knowledge all play a part in determining how well they will perform, as does their search experience and gender. In addition, both the search engine they use and the nature of the task to be performed affects children's performance. Lazonder *et al.* (2000) found that the effect of factors such as these was evident at different stages of searching, with experience and knowledge related to search engines affecting the ability to locate appropriate sites but not information within the sites.

From these comments, the potential of the Internet is clear, but it is also clear that students need to use it, and the information it provides, properly and effectively.

Reports from overseas suggest that students, and in many cases adults, do not search the Internet as effectively as they could. Elliott (2000) commented that: 'My first mistake was to assume that students were experts in the field of Internet research. Watching them cruise through the Internet with speed and what looked like skill, and listening to them tout themselves as experts, I was taken in: it certainly looked and sounded like they knew what they were doing. However, I later realized that, while fast and experienced with Internet searches, they were not actually skillful' (p. 88).

Skills and Knowledge Necessary for Successful Searching

Previous research exploring children's search behaviour on online library catalogues (e.g. Borgman, 1996; Solomon, 1993) has divided the necessary skills or knowledge in different ways. Borgman (1996), for example, identified three areas of knowledge or skills that were necessary for effective catalogue searching:

- conceptual knowledge of the information-retrieval process: translating an information need into a searchable query
- semantic knowledge of how to implement a query in a given system: the how and when to use system features
- technical skills in executing the query: basic computing skills and the syntax of entering queries as specific search statements (p. 495).

More recent research into children's searching using the Internet has identified a number of other types of knowledge that are necessary for successful searching using this media, and they can be divided into three categories of skills or knowledge: search skills/knowledge of process; system skills/knowledge of system; and general skills/knowledge.

Search skills/knowledge of process

This factor encompasses the skills and knowledge that are necessary to search for information using any media. As such, it includes the ability to identify research questions (e.g. Schacter *et al.,* 1998), and to formulate queries and conduct searches, revising where and as necessary (e.g. Bilal, 2001; Large & Beheshti, 2000). It also encompasses the knowledge necessary to synthesise and evaluate the information found, taking into account such factors as the author of the work and their credibility (e.g. Schacter *et al.,* 1998).

Information skills have been identified as one of the essential skills in *The New Zealand Curriculum Framework* document. It is expected that these skills will not be learned in isolation, but rather 'will be developed through the essential learning areas and in different contexts across the curriculum' (Ministry of Education, 1993, p. 17). In the area of information skills, it is expected that students will:

- identify, locate, gather, store, retrieve, and process information from a range of sources;
- organise, analyse, synthesise, evaluate, and use information;

- present information clearly, logically, concisely, and accurately;
- identify, describe, and interpret different points of view, and distinguish fact from opinion; and
- use a range of information-retrieval and information-processing technologies confidently and competently (p. 18).

Specific skills/knowledge of system

Searching the Internet requires specific skills and knowledge. Students must understand how the different search engines work, so that they can make a good choice of which search engine or engines to use, and can also make effective use of them (e.g. Bilal, 2001, 2002). This includes knowledge such as how to use Boolean logic, which search engines support natural language and what punctuation should or should not be used (e.g. Bilal, 2000; Bilal & Watson, 1998).

General skills/general knowledge

In order to conduct searches effectively, students have to know how to use a computer. If they choose to use keyword searching they also need to spell keywords correctly, and to have the reading ability to read and understand the results (e.g. Bilal & Watson, 1998; Fidel *et al.,* 1999). They may also need knowledge of the area in which they are looking for information, so that they know what words will be effective keywords, or in which subject category they should look (e.g. Bilal, 2000, 2001, 2002). At least some knowledge of the topic will also aid them in determining which of the sites presented in the results page are likely to give them the information they need.

Eagleton and Guinee (2002) categorised the skills needed for successful and effective Internet searches in a different way, breaking them down into five categories:

- understanding the inquiry task;
- understanding categories;
- understanding the information space;
- understanding keywords; and
- understanding search engines (p. 40).

The first two categories apply to all forms of searching, with the final three being specific to using the Internet for searching. Their first skill, understanding the inquiry task, is the equivalent of the search skills/knowledge of process category described above; however, they note that the sheer size of the Internet means that several aspects of this are particularly difficult when using this medium. The second skill identified by Eagleton and Guinee, understanding categories, is covered in the general skills/ general knowledge category. Like Bilal (2000, 2001, 2002) the authors recognise the importance of having knowledge of the topic of interest, to enable students to formulate appropriate and effective search strategies.

Eagleton and Guinee's (2002) final three categories are specifically related to the use of the Internet for searching, and correspond to the specific skills/knowledge of system category above. They note that when using the Internet for searching, students must be able to translate their thoughts on the topic of interest into coherent keywords as 'unlike a librarian, a computer takes the students' request for information at face value and is not flexible enough to interpret clumsy queries' (p. 41). It is also important that students understand how the various search engines work, so that they are able to choose the one most appropriate for their search, and then use it to its potential. Finally, Eagleton and Guinee recognise that children need to understand what the Internet is, and how it works. It is important that children realise that, unlike in a library, the information on the Internet has not necessarily undergone a 'professional review and editing process' (p. 41). The size and nature of the Internet also makes it likely that children will not find one site that contains all the information they need, but rather that they will have to gather information from a number of different sites. This is in contrast to traditional sources of information, as 'in a library, students can often locate single books that are complete accounts of their topics' (p. 41).

Many teachers overcome the difficulties associated with students searching the Internet for information by providing them with a list of appropriate websites. Eagleton and Guinee (2000) believe that although this protects students from inappropriate sites and saves time searching, it does not benefit the students. They note that this may prevent students from learning the skills necessary to search the Internet, which will disadvantage them in the long term. As Eagleton and Guinee (2002) say, 'focusing on the inquiry process supports the overarching goal of teaching students strategies they can use in multiple contexts' (p. 40). There are many examples of ways in which students can be encouraged to learn how to effectively use the Internet, with a number described in the appendices.

Conclusion

Access to the Internet, and the wide variety of information contained on it, is seen as one of the benefits of ICT, and research is one of the most common uses of ICT in the classroom. Despite this, however, children's ability to search for information on the Internet is generally poor, and shows a limited understanding of what the Internet is, how it works, and how best to search for information on it. In addition, some children see searching the Internet as 'research' and fail to understand that research involves more than locating information that may be relevant to the topic at hand. Despite these difficulties, it is likely that the Internet will continue to grow, and that children will continue to use it, meaning it is imperative that they learn how to use it effectively.

Appendix A: What is the World Wide Web?

The World Wide Web (Web) refers to all the publicly accessible files (pages) on the Internet that are linked to each other. These files, which may contain text, graphics, audio, or video information, are located on websites, each with a unique address (URL). Currently it is estimated that there are about three billion documents available on the Web (http://www.lib.berkeley.edu/TeachingLib/Guides/Internet/ThingsToKnow.html), but there is no comprehensive index of the files or information that are available. To find information on the Web, you can either enter the URL of a website using a browser software (eg, Netscape or Internet Explorer) that may contain the information you need, or you can undertake a 'search' for that information. It is generally accepted that there are two ways of searching the Web, either using a subject directory based search (eg, Yahoo) or a keyword based search engine (eg, Google).

Subject directories are databases consisting of human-selected websites (pages) organised into categories and sub-categories. In searching these directories, users select the category in which they are interested, or conduct searches using general terms, with the subject categories and description being searched. Examples of directories include: About, Academic Info, Ask Jeeves, BBCi, DMOZ/Open Directory, Infomine, Internet Public Library, Librarians' Index Starting Point, Virtual Library and Yahoo.

Search engines allow the user to search much larger databases of Web pages compiled by 'spiders', 'bots', or 'crawlers' (computer programmes) with minimal human oversight. Searchers use keywords to match the words in the Web pages. Meta-search engines search several of the databases of individual search engines at once, usually presenting the first results from each of these individual search engines. Examples of search engines include: Alta Vista, AlltheWeb, Dogpile, Google, Hot Bot, Lycos and Northern Light; while Metacrawler and Copernic are examples of meta-search engines.

Search engines present their results using a ranking system. The criteria search engines use for page ranking are varied, and the emphasis placed on each varies. Some of these criteria include:

- the number of times the keyword(s) appears in the body of the text on the page;
- whether keywords are in the title;
- whether keywords are in the description metatag (the information that may appear on a results page below the title);
- link popularity (number of external pages which link to this page);
- the location and frequency of the appearance of the keywords on the page;
- whether any penalties/exclusion for 'spamming' (deliberate attempts to increase rating) are enforced; and
- taking into account which pages have been entered by previous searchers using the same keywords.

Table 1 highlights some of the differences between search engines and subject directories.

The distinction between the two types of searches is becoming increasingly blurred. Many subject directories now also offer a keyword search option. For example, Yahoo, which originally offering only directory-based searches now gives users a choice of using directory-based or keyword searches. Similarly, some search engines now have a directory based search available. For example, Google has a directory-based search available that has resulted from combining Open Directory

Table 1: Comparison of subject directories and search engines.

Directory	Search engine
• smaller index • compiled by humans • searches descriptions • organised into categories • generally produces more targeted results	• larger index • compiled by computer programmes • searches full text • ranked by criteria determined by individual search engine

with Google's page ranking system. It is important to note that no one search engine gives access to all the pages on the Internet, with each including only a subset. How much of the Web each search engine gives access to depends on which search engine is used.

The way in which searchers undertake the search process, the types of searches they can undertake, and the way in which the results are presented all differ depending on the type of search engine, as well as between individual search engines. In addition to offering different options for searching, search providers also differ in the way they deal with issues such as plurals, capital letters and the use of natural language. Most search providers support the use of at least the most common Boolean operators or commands such as AND, OR and NOT. These commands were first used in database queries, but also have the potential to be of use in searching for information on the Internet, with the AND, OR and NOT commands being most commonly supported by search engines. In addition to these Boolean operators, many search providers also support the use of other functions at the key word entry stage. The following figure gives details of variations used at the keyword stage, and descriptions of how they may be treated. These variations are based on those used by students searching for information on the kiwi bird (see Lai & Pratt, 2004).

Figure 2. Common functions used by search engines at the keyword entry stage.

Function	Explanation
Plurals and other common endings	
kiwis singing	Some search providers truncate common endings, e.g., removing the letter 's' or the letters 'ing' if they appear at the end of words. Generally, however, they rely on users using wildcards (see separate entry below) to indicate when they wish the search to be conducted on the stem.

Function	Explanation
Capital letters	
Kiwi KIWI kiwi	Most search engines will treat lower case letters as either lower or upper case (so entering kiwi will return KIWI, Kiwi and kiwi) but treat upper case as they are entered (so entering Kiwi will only return Kiwi, and KIWI will only return KIWI). Others ignore case completely.
The use of natural language	
How did the kiwi get its name?	Some search engines ignore common words such as the. In the query shown here, *Google* ignored the words 'how' and 'the' when conducting its search.
Boolean operator: AND or +[#]	
kiwi AND bird kiwi + bird	Searches for pages that contain the word kiwi AND the word bird. Results in 66,400 hits[##]. Most search engines use this command by default, so entering 'kiwi bird' is the equivalent as entering 'kiwi AND bird'.
Boolean operator: OR or |[#]	
kiwi OR bird kiwi |bird	Searches for pages that contain the word kiwi OR the word bird. Results in 21,700,000 hits[##]. Most search engines support this.
Boolean operator: NOT or -[#]	
kiwi NOT person kiwi -person	Searches for pages that contain the word kiwi but NOT the word person. Not all search engines support this.
Boolean operator: NEAR	
kiwi NEAR bird	Searches for pages that contain the word kiwi NEAR the word bird. Different search engines specify different word limits within which the words must appear. For example, *Alta Vista* interprets near as being within 10 words, *Lycos* within 25

[#] whether the word in uppercase (e.g., AND), the word in lower case (e.g., and) or the symbol (e.g., +)must be used depends on the individual search engine
[##] using Google to conduct the search in November 2003

Function	Explanation
	words, and *AOL* as being next to each other, unless the number of words is specified (NEAR/# where # is the number of words).
	Not all search engines support this.

Boolean operator: BEFORE

kiwi BEFORE bird	Similar to the near search, however the word kiwi must appear BEFORE the word bird.
	Different search engines specify different word limits within which the words must appear.
	Not all search engines support this.

Boolean operator: AFTER

kiwi AFTER bird	Similar to the near search, however the word kiwi must appear AFTER the word bird.
	Different search engines specify different word limits within which the words must appear.
	Not all search engines support this.

PHRASES

| "kiwi bird" | Searches for pages that contain the exact phrase enclosed in quotation marks. |
| | Most search engines support this. |

WILD CARDS

| kiwi* | Searches for words that begin with the exact letters prior to the asterisk, with any letter or combination of letters where the asterisk is. In this example, kiwis and kiwi would both be found. |
| | Some search engines support this. |

STEMMING

| kiwi | Related to wild cards, with the only difference being you don't need to add the asterisk, so in this example kiwi and kiwis would both be found. |
| | Some search engines support this. |

PARANTHESES or NESTING

| kiwi AND (bird \| animal) | The use of parentheses to indicate which command or set of commands should be dealt with first. |
| | Some search engines support this. |

Most search engines also have an 'advanced features' option. This may provide searches for words specifically in the title, URL, host, domain or links. It may also offer features such as:

- related searches;
- clustering;
- find similar;
- search with results;
- specify language;
- page translation;
- port filter/warning;
- specify file format; or
- specify when last updated.

Some search engines also offer users the opportunity to alter or adapt the way results are presented in the advanced features option, for example, by altering the number of listings, requiring the date to be displayed, displaying only the titles or sorting the results by date.

Appendix B: Examples of Activities Used to Help Students Understand how to Effectively use the Web

Teaching Boolean logic

Elliott (2000) describes a task created by Laura Cocozzella and Kearney Francis of Montgomery County Public Schools that assists students in understanding Boolean logic and how it may help them find information on the Internet. In this task students become Internet sites, while the teacher is the searcher. As such, the teacher calls out descriptors, or keywords, and if students fit the descriptor, they must stand up. Elliott describes how this might work:

> Of course, you can start very broadly and call out, 'Student', having the entire class stand up. You might react by saying, 'Oh my, there are so many of you that I could not possibly talk to you all about the information that I need. I will have to narrow my search'. Then call out, 'Student AND "movie fan"'. My guess is that only a few students would sit down. You might reinforce the idea that there are still too many 'sites' of information for you to deal with, so you're going to narrow and refine your search further. Next, call out, 'Student AND "romantic movie fan"'. At this point, some of the students will probably sit down. Then call out, 'Student AND "romantic movie fan" AND "Titanic fan"'. Some more students might sit down. For a final narrowing, call out, 'Student AND "romantic movie fan" AND "Titanic fan"' NOT 'Leonardo DiCaprio fan'. By this point, you should have only a few people left standing. Then, you could say to the students, 'Okay, you are the exact people with whom I want to talk!' You can use all kinds of different categories for this activity from sports, to TV shows, to cars, to show students not only how a search engine actually works, but also to show them the power of Boolean search strategies, such

as adding the Boolean commands AND, NOT, OR, or quotations around phrases. Having them enact the part of a search engine might give them a better understanding of what they are doing when they search the World Wide Web and how Boolean tools can help them in their search (p. 88).

Activities that help students learn to search the Internet

Eagleton and Guinee (2002) describe two activities that help students learn how best to search the Internet:

Keyword category maps

Students are required to find keywords into teacher-prepared concept maps. They note that students need to practise coming up with different kinds of keywords, including:

- keywords that are directly stated in the question;
- keywords that need to be substituted from the question;
- keywords from searches requiring more than one keyword.

They gave the example of a hero project with eighth-grade students, using the following concept map:

Figure 3: Eighth-grade hero project keyword-category map (adapted from Eagleton & Guinee, 2002, p. 46).

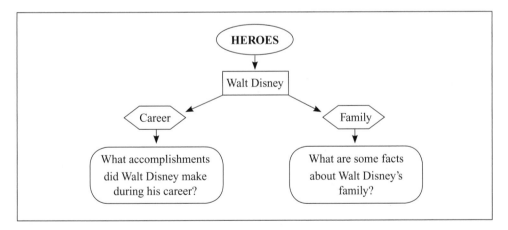

The student used this to come up with two search terms: "Walt Disney" + career and "Walt Disney" + family for her research.

Scavenger hunts

These involve getting students to search for answers to specific questions on a variety of topics, recording search engines and keywords used. This enables teachers to identify how proficient students are at searching, as well as facilitating class discussion as to why certain search engines and/or keywords were more or less successful. They suggest either linking to existing scavenger hunts (e.g., http://www.yahooligans.com/tg/basil. html) or creating your own questions. Examples of questions they've used with middle school students were:

- Which US President got stuck in his bathtub on inauguration day?
- Who was the inventor of basketball?
- Who discovered King Tut's tomb?
- What is the name of North Carolina's professional women's soccer team?
- Who was the first African American to win the Nobel Prize for Literature? (p. 46)

Internet information literacy unit

O'Sullivan and Scott's (2000) three-day unit was designed to be used with social studies classes, but can be adapted for subject area.

Day one:
- Teach students "the structure and terminology associated with the Internet, how to translate and dissect a URL, and present examples of biased and unreliable Web sites" (p. 41).
- Introduce a five-step evaluation criteria (based on that developed by Tate & Alexander, see below) and use this to evaluate three different Internet sites that deal with one topic.

Days two and three:
- Students select a subject from a list of commonly studied social science topics and evaluate the first three Web sites listed using the evaluation criteria.

Web Site Evaluation Worksheet
Topic: Search Engine:
# of hits: URL

1. *Accuracy* (Is the information reliable? Are the links accurate? Sources cited? Information believable?)

2. *Authority* (Who is the author of the site? What are his/her qualifications? Is the site sponsored by an organisation? Is the organisation reputable or legitimate?)

3. *Objectivity* (Does the information reveal a bias? What is the point of view of the author? Is the information trying to sway you? Do the links also reflect a bias?)

4. *Currency* (When was the site last updated? Is the information kept up to date? Is the publication date indicated? Are the links up to date?)

5. *Coverage* (How is the information presented? Heavy use of graphics, text, statistics? Topic coverage cursory of in-depth?)

Checklist for an Informational Web Page

The above worksheet is based on Tate and Alexander's work on finding 'informational' Web pages which is available at http://www2.widener.edu/Wolfgram-Memorial-Library/webevaluation/inform.htm). This and other checklists developed to help aid students to evaluate Web Resources (http://www2.widener.edu/Wolfgram-Memorial-Library/webevaluation/webeval.htm) are able to be reproduced in their entirety for educational purposes.

References

Becker, H.J. (1999). *Internet use by teachers: Conditions of professional and teacher directed student use. Report #1.* Center for Research on Information Technology and Organisations, University of California, Irvine and University of Minnesota (February). Retrieved 16 February 2002: <http://www.crito.uci.edu/TLC/findings/interne>.

Bilal, D. (2000). Children's use of the Yahooligans! Web search engine. I. Cognitive, physical and affective behaviors on fact-based search tasks. *Journal of the American Society for Information Science, 51*(7). 646–65.

Bilal, D. (2001). Children's use of the Yahooligans! Web search engine: II. Cognitive, physical behaviors on research tasks. *Journal of the American Society for Information Science, 52*(2), 118–36.

Bilal, D. (2002). Children's use of the Yahooligans! Web search engine: III. Cognitive, physical behaviors on fully self-generated search tasks. *Journal of the American Society for Information Science, 53*(13), 1170–83.

Bilal, D. & Watson, J.S. (1998). *Children's paperless projects: Inspiring research via the web.* Paper presented at the 64th IFLA General Conference, 16–21 August 1998, Amsterdam. Retrieved 5 September 2003 from: <http://www.ifla.org?IV/ifla64/009-131e.htm>.

Borgman, C.L. (1996). Why are online catalogs *still* hard to use? *Journal of the American Society for Information Science, 47*(7), 493–503.

Bruce, B. (1999/2000). Searching the Web: New domains for inquiry. *Journal of Adolescent & Adult Literacy, 43*(4). 348–54.

Eagleton, M.B. & Guinee, K. (2002). Strategies for supporting student Internet inquiry. *New England Reading Association Journal, 38*(2), 39–47.

Elliott, C.B. (2000). Helping students weave their way through the World Wide Web. *English Journal, 90*(2), 87–92.

Fidel, R., Davies, R.K., Douglass, M. H., *et al.* (1999). A visit to the information mall: Web searching behavior of high school students. *Journal of the American Society for Information Science, 50*(1), 24–37.

Ford, N., Miller, D. & Moss, N. (2001). The role of individual differences in Internet searching: An empirical study. *Journal of the American Society for Information Science and Technology, 52*(12), 1049–66.

Jansen, B.J. & Pooch, U. (2001). A review of web searching studies and a framework for future research. *Journal of the American Society for Information Science and Technology, 52*(3), 235–46.

Lai, K.W. & Pratt, K. (2003). *Learning with technology II: Evaluation of the Otago Primary Schools Technology Project.* Dunedin: Community of Otago Trust.

Lai, K.W. & Pratt, K. (2004). Evaluation of the NEMP compact model of the World Wide Web for exploring children's information skills. Report submitted to the NEMP group.

Lai, K.W., Pratt, K. & Trewern, A. (2001). *Learning with technology: Evaluation of the Otago secondary schools technology project.* Dunedin: Community Trust of Otago.

Lai, K.W., Pratt, K. & Trewern, A. (2002). *Use of information and communication technology in Otago secondary schools: Phase 1 of a three-year study.* Dunedin: Community Trust of Otago.

Large, A. & Beheshti, J. (2000). The Web as a classroom resource: Reactions from the users. *Journal of the American Society for Information Science, 51*(12), 1069–80.

Lazonder, A.W. (2000). Exploring novice users' training needs in searching information on the WWW. *Journal of Computer Assisted Learning, 16*, 326–35.

Lazonder, A.W., Biemans, H.J.A. & Wopereis, I.G.J.H. (2000). Differences between novice and experienced users in searching information on the World Wide Web. *Journal of the American Society for Information Science, 51*(6), 576–81.

Market Data Retrieval (1999). *Technology in Education, 1999.* Retrieved 21 September 2000, from the World Wide Web: *<http://www.schooldata.compr18.htm>.*

Ministry of Education (1993). *The New Zealand Curriculum Framework.* Wellington: Ministry of Education.

Nachmias, R. & Gilad, A. (2002). Needle in a hyperstack: Searching for information on the World Wide Web. *Journal of Research on Technology in Education, 34*(4), 475–86.

O'Sullivan, M. & Scott, T. (2000). Teaching Internet information literacy: A critical evaluation. *Multimedia schools, 7*(2), 40–2, 44.

Palmquist, R.Z. & Kim, K.S. (2000). Cognitive style and on-line database search experience as predictors of web search performance. *Journal of the American Society for Information Science, 51*(6), 558–66.

Schacter, J., Chung, G.K.W.K. & Dorr, A. (1998). Children's Internet searching on complex problems: Performance and process analyses. *Journal of the American Society for Information Science, 49*(9), 840–9.

Smerdon, B., Cronen, S., Lanahan, L. *et al.* (2000). *Teachers' Tools for the 21st Century.* The National Center for Education Statistics: Statistical Analysis Report. Retrieved 16 June 2001 from: <http://nces.ed.gov/pubsearch/pubsinfo.asp?pubid=2000102>.

Solomon, P. (1993). Children's information retrieval behavior: A case analysis of an OPAC. *Journal of the American Society for Information Science, 44*(5), 245–64.

Sullivan, C. & Anso, M. (March, 2000). *ICT in schools 1999.* Report prepared for Information Technology Advisory Group (ITAG). Retrieved 30 August 2000, from: <http://www.med.govt.nz/pbt/infotech/ictschools1999/ictschools1999/html>.

Wallace, R. and Kupperman, J. (1997). On-line Search in the Science Classroom: Benefits and Possibilities. In Soloway, E. (Symposium chair). Using On-line digital resources to support sustained inquiry learning in K-12 science. Paper presented at the AERA, Chicago, 1997. Retrieved 9 October 2002, from: <http://www.msu.edu/~ravenmw/pubs/online_search.pdf>.

11

Cybersafety: An Intrinsic Part of the Online Experience

Liz Butterfield

Introduction

Communication technologies are making fundamental changes in the lives of children and young people. The Internet and mobile phones are now merging, creating new, more portable devices that easily access the virtual, digital world of cyberspace. All of these technologies offer incredible learning opportunities, as well as some serious challenges regarding safety and security in the school, in the workplace and in the home environment.

Those who introduce children to these technologies and foster the many e-learning opportunities the Internet offers have a very real obligation to educate about safe and responsible behaviour in cyberspace. Just as 'bricks-and-mortar' schools have a duty to provide a safe physical learning environment, teachers in virtual classrooms have a similar responsibility to prepare children and young people for their online experience and empower them with the knowledge that can help keep them safe.

Addressing the real risks associated with technologies should not be dismissed as overly reactive or pessimistic. When considering the vast and ever-growing amount of course work that many educators are giving to students, one seldom if ever sees any mention of online safety concerns. This is analogous to handing a student, who has never had a driving lesson, the key to a high-power automobile, and enthusiastically pointing them in the direction of the eight-lane superhighway. Our use of these technologies in innovative classroom work has matured to the point where an absence of safety education is simply no longer acceptable. Even students who are experienced users of the Internet are not necessarily knowledgeable about safety issues. Periodic review of salient points about safe and responsible use, while keeping pace with new developments in cybersafety, is very important.

Risks and Challenges: The Reality of Cyberspace

The first section of this chapter will discuss the difficult and complex situations one can encounter online and a later section will address the constructive and effective ways to minimise such risks.

The dimensions of cyberspace

It is easy to forget the number of people who could potentially link to one's personal information or communications in cyberspace. The paradigm shift from pre-Internet 'community access' to the 'global access' of the Web creates wonderful opportunities, but also serious risks. The 14-year-old New Zealand girl who created her own website on which she posted a map of where she lived, or the schools that have posted individual student photos with the students' names (and without parental permission) are good examples of how tech-savvy Internet users can forget about the possible ramifications of publishing in cyberspace. Because of powerful search engines and the growing use of 'bots' to roam the Internet and harvest particular types of information, one has to assume anything put on the 'net' can potentially be accessed by millions of people. Yet it is hard to conceptualise the massive scale of this medium of communication.

Privacy and anonymity

The concept of a right to privacy is being seriously challenged in cyberspace. Commercial websites are not required to post privacy statements informing visitors of how much personal information is being collected and what is done with that information. For example, a person who shops at an online bookstore typically has a significant amount of information recorded about who they are and their shopping profile. Yet visitors to such websites may be completely ignorant of the use of 'cookies'. Small information-gathering programmes called 'cookies' can covertly record details about you, such as who you are, what you are doing, and where you have been, to be used in future transactions or for market research. Many like the convenience of having the cookie used to customise web pages on their return to a website, but others might object. Third parties, often advertisers on sites visited, can also attach cookies to customise ads displayed on future visits. The next step beyond cookies is spyware, where, for example, free software downloaded off the Internet contains a hidden programme that surreptitiously reports back to the manufacturer about a person's profile or preferences. It is very hard for individuals to control their own personal information in this environment.

There are many good reasons to try and maintain anonymity and privacy in cyberspace, and some nefarious ones as well. There exist some effective mechanisms for making one's identity hard to trace. Pre-pay mobile phones can be purchased relatively cheaply by young people who want to keep their phone activity under parental radar; such pre-pay phones can also be easily discarded after being used for criminal purposes. On the Internet, some businesses offer anonymous re-mailing services that can make a communication more difficult to trace, and more of a challenge to law enforcement.

The balance between personal privacy in cyberspace and a society's need to protect its citizens, especially children, from e-criminals who would try and cloak themselves in anonymity will be the subject of rigorous debate in the coming years.

Disinhibition

Whatever the reality, many *perceive* the Internet to be a private and an anonymous space, which can contribute to 'disinhibited' behaviour. There is a growing body of research that indicates that many people may behave in cyberspace in a way they would not normally – that they easily lose their inhibitions (see Sproull & Kiesler, 1986; Reid, 1998; Joinson,1998). Other contributing factors may be that people often use the Internet in the comfort of their own homes, and that cyberspace interactions are initially less awkward and uncomfortable than meeting new people face to face.

Disinhibited behaviour can be very positive – for example, with people being more outgoing or more confident online (which can have a wonderful impact on e-learning). However, disinhibition may also manifest in people taking greater risks on the Internet, confiding personal information much more quickly than in face-to-face interactions. Online relationships can accelerate very rapidly to a level of trust, a factor expertly exploited by sexual predators. Disinhibition can also lead to acting in a more aggressive, anti-social manner online than offline. In addition, forgetting the scale of the Internet and the speed with which material can widely circulate can contribute to impulsivity. Setting up a website to cyberbully someone or cruelly circulating very private information, such as the highly sexual 'love letter' that was widely circulated around the globe in a matter of hours, are examples of such mean-spirited behaviour.

Raising awareness of disinhibition and encouraging young people to moderate their own online behaviour will become more important as hand-held technologies (mobile phones, PDAs, and so on) become more popular and affordable, giving 24/7 access to the Internet with little opportunity for parental (or teacher) monitoring.

Chat environments
Chat environments such as instant messaging (IM), mobile phone text chat, instant relay chat (IRC), ICQ and other chat venues allow conversations in real time with people all over the world. These conversations may be with a group or with a single person, and may be with people known previously or with those met online. This is where children and young people are at greatest risk on the Internet because a person can easily assume any persona they want. Traditional 'stranger-danger' safety messages regarding the Internet are very ineffective, because strangers are so quickly perceived as trusted friends. This is also why Microsoft recently announced the closing of free, user-created MSN chatrooms: they were too unsafe for children.

The content of many chatroom interactions can heighten the risk. Many young people engage in conversations that involve flirting or sexual language. In a survey done by the Cyberspace Research Unit in Britain, 40 per cent reported engaging in conversations of a sexual nature with online 'friends' (O'Connell, 2003a). Adolescents would have a natural curiosity about experimenting with sexual language and behaviour online in a 'cybersex' conversation, but the person they are having this highly sexual discussion with may be much older and/or have a very different agenda.

The risk level increases further when a face-to-face meeting is set up with an online acquaintance. In the Girls on the Net New Zealand survey, 34 per cent of the 11- to 19-year-old respondents met someone face-to-face that they had met online (Internet Safety Group, 2001). In The Net Generation New Zealand survey (2500+ respondents), 23 per cent of seven to ten year olds who used the Internet, and 37 per cent of those over sixteen, reported going to such face-to-face meetings (Internet Safety Group, 2002). It is worth noting that most of these interactions were positive, with new 'friends' reported as being close to the age they had said they were. Interestingly, in this survey of students from a comparatively low-income area (decile three and four), 46 per cent of the eleven to nineteen year olds had their own mobile phones and 25 per cent were using those phones to interact with people they didn't know (Internet Safety Group, 2002).

Safety strategies are absolutely essential when setting up such face-to-face meetings. In the Girls on the Net survey, 32 per cent of respondents went to that meeting alone (Internet Safety Group, 2001). In the Net Generation survey, the students were asked what their safety rules were for meeting people they didn't know well and only 37 per cent reported they would tell their parents, and 18 per cent reported they had no safety rules at all (Internet Safety Group, 2002).

Our society does not have a widely recognised code of social behaviour yet for online relationships, but with effective education that can quickly evolve. Then young people will know instinctively at what age they need to take a trusted adult with them to such a meeting and when they can go just with their friends. They will know to set the meeting up at a public place, have transportation planned to leave if they are uncomfortable and to keep things fairly anonymous until that first encounter. Also, they will know that behaviours from a new online friend, such as pressuring to come to a meeting alone, are a clear warning sign.

Internet relationships are here to stay. Young people are using cyberspace not just for maintaining existing relationships, but also for seeking out new acquaintances: usually for friendship, perhaps for romance and sometimes for sexual interaction. One of the great strengths of the Internet is the ability to connect with people all over the world and learn of different cultures and ideas, as well as connect with friends and family almost anytime. Thus, safety messages like 'don't give out personal information' or 'don't meet people face-to-face that you meet online' are obviously irrelevant to the many who view the Internet and the mobile phone as intrinsic elements of their social milieu and who quickly feel their interactions are with trusted friends.

Sexual predators

The number of sexual assaults of children and young people by online predators, both in New Zealand and overseas, underscores how serious the risk online can be. Sexual predators are adept at exploiting people's trust, so will look for environments where people are more likely to let their guard down. We also know that most paedophiles, both adolescent and adult, often deliberately select children for their availability and vulnerability.

There is often a process of preparing a child for abuse known as 'grooming'. Ilene Berson (2002), from the University of South Florida, presented a paper at the 2002 NetSafe symposium on this topic.

Grooming is also a deceptive process in which a child is unprepared to interpret cues which signal danger of risk. Predators are skilled at gaining the trust of a child before luring them into interactions. The process of grooming through the formation of a close bond creates a victim who is more likely to comply with sexual advances.

Our interpretations of the grooming process have been primarily anecdotal, based on the accounts of child victims and predators who might recall gifts given, and tactics used to gain trust and establish confidence. In addition to these narrative accounts, evidence might be found of adult or child erotica, photographs of children, photography equipment, or children's play items. However, grooming evidence is now more available in the form of chat logs, stored communication and downloaded child and adult pornography, gifts, notes about online conversations, and profiles on youth met online (p. ?).

There are several key differences between online and face-to-face grooming.

On the Internet, child molesters are able to shorten the trust building period and establish trust with many more potential victims at once. Because the Internet is anonymous, child molesters can pretend to be virtually anyone in order to deceive the victim into thinking that they are understanding and sympathetic. One good example is from a case in which the defendant pretended to be the same age as his victim. Like his victim, the defendant said that his parents were recently divorced and he was having trouble adjusting to his new life. After a certain period, the defendant said his father 'banned' him from the Internet, but his older brother would be able to continue the Internet friendship. Thus, over time, the victim not only came to trust the defendant, but gradually accepted the fact that the defendant was older than him and an adult (Brown, 2001).

In online grooming, a perpetrator may carefully lead a child into sexual conversations:

The nature of sexual conversation will vary from mild suggestions to explicit descriptions of, for example, oral sex. The focus may be on the child, i.e. the adult asking the child to touch itself and to explain what it feels like. The usual rationale for this approach is that the adult is somehow perceived as a mentor who will guide the child to a greater understanding of his or her own sexuality ... Research findings indicate that this pattern of conversation is characteristic of an online relationship that may progress to a request for a face to face meeting ... (O'Connell, 2003, p. 10).

The easy access to children and young people that the Internet and mobile phones offer to sexual predators must be countered with education about the risk. This situation is also presenting challenges for sexual-offender treatment programmes. How these programmes ascertain the true extent of an offender's Internet use, and how that use is monitored, will be important questions to address in sexual-offender treatment.

Online use of cameras
'Web cams' can play a vital role in distance learning, but without education about responsible use this helpful technology can easily put children and young people at risk. Web cams allow participants in 'real-time' online conversations to view one another. Although the camera allows one to easily verify the age and gender of a fellow chatter, the web-cam image of a young person aids in their identification, as well as sometimes offering further identifying information in the background, such as town, school, sports teams, and so on.

One case reported to the ISG involved a young person using a web cam as requested, but the 'online friend' had a steady stream of excuses why his or her web cam wouldn't work. Also reported have been cases where young people in online relationships with paedophiles have been asked to appear in sexually explicit poses before the camera (essentially creating child pornography) or where a young person has sent inappropriate images to a peer (with whom that young person has an intimate relationship) and the images were then widely distributed by the recipient.

Helping young people understand, again, the scale of the Internet, especially the inability to control the publication of images once they are online, can help. Also, simple safety strategies such as unplugging the camera after use can prevent a web cam (often with an excellent view of a young person's bedroom) being covertly activated by a hacker.

Small mobile phone cameras have made taking, sending, and posting photos over the Internet effortless. However, the small size of the phone camera makes it easy to photograph someone surreptitiously and quickly post on the Internet very private/personal images (potentially from a school locker room, gym, or public pool). Here both awareness that one's photo *can* be secretly taken with these phones *and* that responsible use would dictate asking someone's permission before taking a photo can be constructive. Companies that manufacture handsets can help by ensuring that a flash of light and a sound happen when a picture is taken. These should be standard phone features that cannot be disabled.

Harassment – from cyberbullying to death threats

There is a broad continuum of harassing behaviour that can be perpetrated using communication technologies. On the one end is classic schoolyard bullying. In cyberspace, such bullying has additional impact because the victim cannot go home and feel safe; as well, the apparent anonymity of the messages may add to the victim's feeling of helplessness. If done within a school community, bullies are usually fairly easy to identify, yet victims of cyberbullying may feel the messages are untraceable.

Cyberbullying can take many forms: emails, instant messages, text messages, 'spoofing' the victim's identity by using his or her e-mail address, circulating the victim's contact details or other personal information, putting derogatory or obscene messages about the victim in chatrooms or on message boards, creating websites about the victim and inviting often obscene 'guest-book' comments. Messages can also be sent repeatedly, as with one New Zealand student who recently received 14,000 copies of the same e-mail, which shut down his school's network.

Some cyberbullying moves into the area of civil harassment and/or defamation. If, however, there is a consistent pattern to the harassing behaviour from one individual, and the recipient feels in real danger, then the threshold into criminal harassment may have been crossed and the incidents need to be reported to the police. Such serious behaviour is often labelled 'cyberstalking'. At the most serious end of the harassment continuum are explicit death threats; unlike civil or criminal harassment, these can be one-off incidents and still amount to a criminal offence. Several incidents reported to the ISG in 2003 that happened in New Zealand schools involved explicit death threats.

When looking at harassment on a continuum, it is important to remember that cyberbullying can be as upsetting and disturbing to a victim as a death threat, with cases in New Zealand and overseas of suicides by young people in which online bullying was a prominent factor. All activities on the continuum, regardless of legal definition, can be emotionally devastating. Also, behaviour from a perpetrator need not follow a logical progression along the continuum, but can move back and forth between a number of different harassing activities.

The New Zealand Girls on the Net survey results (ages eleven to nineteen) showed that 23 per cent reported feeling unsafe or threatened on the Internet (not all from harassment), yet only 48 per cent would tell a parent, 15 per cent would tell the police and 10 per cent wouldn't tell anyone. In the 'Net Generation' survey (ages seven to nineteen), just 20 per cent of those who had felt unsafe reported telling a parent. There are indications that young people may be less likely to seek help because of a fear that part of their parent's reaction to the problem may be to take away the technology. Also, trusted adults young people might traditionally confide in – teachers, counsellors or parents – are often perceived as having little literacy with these technologies.

In the results of a large US survey (ages ten to nineteen) 6 per cent reported incidents of harassment; 50 per cent disclosed the incident to a parent, while 24 per cent did not tell anyone (Finkelhor *et al.*, 2000). The researchers who conducted the survey offered this interesting conclusion:

> Harassment does not occur as frequently as sexual solicitation or unwanted exposure to sexual material, but it is a problem encountered by a significant group of youth … An important feature of harassment is that, more than sexual solicitation, it involves people known to the youth and people known to live nearby. Certainly, some of the threatening character of these episodes stems from the fact that the targets do not feel completely protected by distance and anonymity. The harasser could actually carry out his or her threats (p. 24).

Results of a survey in Britain showed that one in four children has been bullied or threatened using communication technologies: 16 per cent via mobile phone, while 7 per cent were harassed in chatrooms and 4 per cent by e-mail (NCH, 2002).

Cybersafety education can obviously play an important role here. Awareness of support options and an understanding that perpetrators can be caught and the hurtful behaviour stopped will help empower victims to come forward. As well, education about the devastating effect such behaviour can have on a victim, and the very real consequences when caught, can illustrate clearly the importance of responsible use and encourage young people to moderate their own behaviour online.

Addiction

As with many technologies, the Internet can be very addictive to some people. As a result a number of experts have even tendered the existence of an Internet addiction disorder (IAD) (Ferris, 2004; Greenfield, 1999). When a young person spends thirty to forty hours a week on the Internet on top of their school time, their social relationships, their school work, even their family relationships may suffer. Things have become out of balance for that young person and they may need some professional support to re-balance the virtual world of the Internet with the real world in which they live.

One of the aspects of Internet activity that young people can become addicted to is online gaming – role-playing games that can be incredibly absorbing and realistic. Several neighbourhoods in Wellington and Auckland, New Zealand, have recently dealt with petty crime waves because young boys were having to finance both their online time at twenty-four-hour Internet cafés and the gambling they were doing on the game outcome.

Scams, fraud and the need for computer security

These are emerging areas of risk online for everyone, but children are less likely to have the financial viability necessary to be defrauded. However, a growing number of students, especially those visiting from overseas, do have their own credit cards and are thus vulnerable. It is worth noting that if a child chooses to misuse their parent's credit card, that parent will be liable for all charges.

How many students could define a firewall or could talk about the last patch they installed on their home computer operating system? Continuing with the car analogy, it is as though we have given young people itineraries for the wonderful trips they can take with the car, but have forgotten to teach them about the need for basic car maintenance. Computer security maintenance must become an integral part of computer ownership because the vulnerable home computer, 'zombie-ised' by a criminal hacker, can be a potent threat. Whether commandeered for a 'denial of service' attack or simply utilised for theft of bandwidth to move large amounts of MP3 (music) or pornography files, the impact on the computer owner can be devastating. Additionally, personal information stored on the hard drive is there for the taking on a compromised computer, which can contribute to the growing crime tsunami of identity theft.

The reality is that in many households the children are the most knowledgeable computer users, sometimes even being the administrators on the filtering packages that are installed. It is also young people who are installing programmes such as peer-to-peer file sharing (usually to facilitate MP3 trading) without any awareness of the serious security implications. So education of young people about the need for (1) a firewall, (2) anti-virus software (3) security updates/patches for the operating system *and* the need for the regular updating of all three, can not only address the immediate problem of vulnerability, but can influence the generations that follow about the need for security in cyberspace.

Online gambling

Gambling is now available twenty-four hours a day online and presents a unique set of concerns such as: problem gambling; gambling while cognitively impaired (e.g. drunk, drugged or mentally unwell), gambling in the workplace and school; difficulty in confirming the identity of the gambler (e.g. are they underage?), and amount of time spent online (Griffiths, 2004).

The technology that allows online identity (or age) verification is in its infancy; thus, many overseas gambling sites are now using credit cards for age verification. This means that minors, using their own or their parent's credit card, can potentially gain entry into an adults-only online casino and gamble to the credit limit of that card. An increasing number of aggressive online casino pop-up ads may increase children's vulnerability to this activity.

Anti-social material

Young people and adults need to be wary of the anti-social information that is readily available on the Internet, including: hate and cult material, drug recipes, weapons and bomb designs and weapon sales sites. Sometimes purveyors of this material display it on very professional, mainstream-looking websites, helping to legitimise whatever particular ideology they are pushing – for example, a pro-anorexia website. Sometimes there can be an anti-social website embedded in another, more mainstream site, to trap the unsuspecting – for example, when accessing information about suicide prevention, one can find that the site has been 'hijacked' by a very racist or homophobic site.

Pornography

Pornography purveyors are some of the most adept marketers in e-commerce. When people either accidentally or deliberately click on a pornography site, multiple windows can open one after another (mouse-trapping), making it very difficult to exit without crashing the computer. Such websites will often automatically load themselves into a user's Favourites folder or onto the user's desktop. Pornographers also register domain names that are one keystroke off the names of popular kids' websites to catch the unskilled typists – for example, two years ago if one entered the address of the New Zealand Ministry of Education's Online Learning website, but forgot to add the '.nz', an Armenian porn site appeared.

Some unscrupulous adult sites have terms and conditions to read before accessing the 'free' pornography, which of course many, especially adolescents, click the 'I agree' button to bypass. How many curious adolescents (or adults for that matter) read terms and conditions? What they may have agreed to is having their online session transferred to an 0900 phone line at a much higher cost. Sometimes, the terms and conditions include downloading software ostensibly to view the pornography. This software is actually an automatic re-dialler programme that can call the 0900 number at regular intervals, resulting in multiple, costly sessions on the phone bill.

Such marketing tactics may have allowed pornography to become a major online industry, but the callous and ruthless exploitation of young people's natural curiosity about sex is a serious concern.

Legal Internet pornography, soon to be available via mobile phones, can be quite extreme in nature and can be found in spam, websites, chatrooms and peer-to-peer networks (as can illegal pornography). We do not yet know the impact of this exposure on children's and young people's sexual development, because such ease of access to such a vast quantity of this sexual material has never been possible before. But a survey in 2002 by the Chinese government showed 70 per cent of young people surveyed were using Internet porn as their sex education (China Daily, 2002). Also, in 2004, the Child At Risk Assessment Unit in Australia released preliminary reports of a 'dramatic' increase in the number of children under ten sexually harming other children. Almost all in their study went online to access pornography and 'many thought that was Internet's sole purpose' (Wallace, 2003). More research in this area is urgently needed.

Illegal 'objectionable' material
Objectionable material, as defined in the Films, Videos and Publications Classification Act 1993, includes a wide range of images, such as bestiality and necrophilia, but the Department of Internal Affairs Censorship Compliance Unit understandably concentrates its efforts on child pornography. In the last four years, over 20 per cent of those caught trading this material by the DIA were school-age males as young as fourteen.

Child pornography is the recording of the horrific sexual abuse or torture of children – usually in still images, but sometimes in videos or films with soundtracks. Whether the subject is infant rape or forced sexual activity between six-year-olds, those moments of abuse are perpetuated by the callous global trading of these images in order to build collections. In some overseas cases, the children did not survive the abuse that was recorded.

This material is most commonly traded in the chat channels – for instance, Internet Relay Chat (IRC). Here the channels are explicitly named, leaving no doubt about what is being offered: 'baby rape' or 'under-two sex pics' are two examples. Such illegal images are also now being offered in the peer-to-peer file sharing environments such as Morpheus and KaZaA. Child pornography can also be encrypted or embedded within other files, such as music files, in order to avoid detection. People who know where to look can find child pornography on the Internet in about ninety seconds.

Many of those who collect child pornography do so by socialising in a virtual community of thousands of like-minded collectors. Young people can be lured into trading with the offer of a free starter collection of images. Currently, money is not involved in most trades, but that too may be changing as child pornography becomes a more lucrative enterprise and organised crime begins to get involved.

For many years, there was a relatively static collection of child pornography world wide. The Internet has now created a huge marketplace for this material and the demand for new images is increasing. Digital cameras make it easier to create and quickly trade new 'homemade' images, which sometimes have been required for admittance into organised paedophile rings.

Child pornography can also be used by some paedophiles to 'groom' children for abuse by normalising such behaviour for the potential child victim. There have been several incidents reported in New Zealand where older paedophiles emailed child pornography images to children, as well as several incidents where young people mailed peers such images.

Recent profiling research was carried out by the Department of Internal Affairs based on observations and interviews with 109 offenders convicted of offences related to objectionable publications on the Internet. Of these offenders, sixteen worked or had frequent contact with children. Sixteen others had access to children that included: babysitting, volunteering at school, coaching a sports team and chatting on the Internet with children (Manch and Wilson, 2003).

Hacking and virus distribution

The Internet presents opportunities for young people to get involved in activities that may seem to them like games or harmless mischief but which could actually have very serious legal consequences. Judge David Harvey (2002) discussed these issues in his paper presented at the 2002 NetSafe symposium:

> It is easy to obtain on the Internet programs or tools which enable access to other computers on the network. These may be described as hacking tools. In addition, many commercially available computer security programs contain within them tools which may be turned to enable unauthorised access to network systems. These programs may be easily used by young people who, sitting in the privacy of their own home, and using the apparent anonymity of the Internet, may access other people's computers and may consider themselves invulnerable. This is simply not the case. (Harvey, 2002)

Young hackers are able to pick up advice, techniques and 'tools' from both chat rooms and hacking sites. Very high profile business and government websites overseas may get as many as 50,000 hacking attempts a day, not all from amateurs.

There are also websites on the Internet to assist with creating viruses, which can then be distributed via email or other means. The new Crimes Act Amendment 2003, and a growing awareness by the public that people who engage in such activity can be reported to law enforcement, will no doubt mean an increase in cases involving hacking and viruses coming before the New Zealand courts.

Copyright infringement

That feeling of invulnerability that Judge Harvey discussed in the previous section is certainly a factor in the trading or copying of MP3 music files. Music industry companies have been rigorous in their monitoring of those who have downloaded their copyrighted material. It remains to be seen how they will use this information about individual culprits. The feeling expressed by so many young people that they are 'too small' for the music industry to bother with may prove to be an unfortunate miscalculation.

A particular area of concern for schools and universities is plagiarism. This can range from the indiscriminate cutting and pasting of others' work into assignments done at home, to the patronising of 'cheating sites', where you can download entire papers and homework assignments for a fee. This surprising variant of e-commerce has, in turn, spawned businesses which, for a fee paid by a university, check over student papers for plagiarism. It seems far simpler to educate young people from an early age about the concept of intellectual property, the reasons for copyright law and the risks of copyright infringement.

Children as perpetrators

One of the risks online for children and young people is becoming involved themselves in anti-social or criminal activity. In New Zealand schools in 2003, there were incidents of criminal hacking and criminal harassment involving children as young as eight. The inclination of many to be more disinhibited online than offline certainly applies to children as well, but there are also serious developmental issues to consider. Children can find themselves in quite complex situations online, which demand a sound and very mature ethical base from which to formulate their responses, yet at the same time they may not fully grasp the potential psychological or financial impact of their actions. As well, the adults children might usually turn to for advice (parents, teachers, and so forth) might be too far behind with the technologies to be of help, or at least may be perceived that way.

By ensuring that cybersafety education stresses both safe *and* responsible use, children and young people can be effectively prepared for what they may encounter online and can better moderate their own behaviour.

Positive Strategies for a Safer Online Environment

When considering the aforementioned risks and challenges, it is heartening to know that there are a number of comprehensive international initiatives actively disseminating education to keep everyone safe in cyberspace. In New Zealand, there is a collaborative programme that educates across all sectors of society, as follows.

New Zealand's NetSafe Programme

The Internet Safety Group (ISG) has developed a comprehensive national programme for the safe integration of communication technologies, including the Internet, into New Zealand society. A primary focus of the ISG is on the safety of children and young people, but the scope of the group's work includes cybersafety education for every New Zealander. The NetSafe Programme is unique in bringing issues such as child safety online and information security for businesses under one 'umbrella'. This holistic approach offers maximum opportunities for cross-sector networking and collaboration.

NetSafe is an educational programme with a clear, consistently positive, commonsense approach. The ISG encourages use of the Internet and mobile phones and consistently emphasises communication and collaboration among organisations, businesses and government bodies.

The ISG includes members of: New Zealand Police, Department of Internal Affairs Censorship Compliance Unit, Customs Service, the judiciary, as well as numerous community groups, educators from primary to university level, students, parents and prominent businesses. The primary sponsor is the Ministry of Education, which has designated the ISG its 'agent of choice' for delivery of Internet safety education. Other sponsors include: New Zealand Police, Westpac, Sanitarium Health Food Company and Multi Serve Education Trust.

NetSafe website and Helpline

The huge growth of www.netsafe.org.nz is where the progress of the ISG is best reflected. The range of topics offers a comprehensive look at cybersafety issues. A growing number of printed resources, covering a wide range of cybersafety topics, can be viewed and ordered from this website. Queries are easily sent from the bottom of every website page, and many visitors have taken advantage of this, including those from overseas. The ISG toll-free phone service in New Zealand, 0508 NETSAFE, has handled many calls from all over the country and from overseas, including queries on cyberbullying, stalking, scams, fraud, copyright infringement, identity spoofing, online auction fraud and more.

National and international conferences

The ISG, the New Zealand Police and the University of Auckland organised an invitation-only national symposium in February 2002, which brought together leaders from a wide variety of different sectors of New Zealand society. An international conference, NetSafe II: Society, Safety & the Internet, building on the same themes as the symposium – safety in the home, business, school and workplace – was held in July 2003 and attracted delegates from all over the world. The proceedings of both events can be accessed on the NetSafe website.

Plans are now underway for an international conference in 2005 organised by the partnership of the ISG, the University of Auckland and the Oxford Internet Institute. This conference, with the same themes as the 2003 conference, will be at Oxford University in Britain.

NetSafe Kit for Schools

In 2003, the NetSafe Kit for Schools was sent to every school and library in New Zealand. This kit built on the success of the ISG's Internet Safety Kit released in 2000. Like its predecessor, the NetSafe Kit guides schools in the establishment of a cybersafe-learning environment. Reports from the Education Review Office (ERO), along with a marked increase in the number of NetSafe interactions with schools, boards of trustees and teachers using the kit, is indicative of both the need for and the acceptance of NetSafe's service.

There are three necessary components to creating and maintaining a cybersafe school learning environment:

- an infrastructure of policies, procedures and signed staff and student use agreements;
- an effective electronic security system; and
- a comprehensive cybersafety education programme for the entire school community.

These components are rigorously enforced by a cybersafety team, led by a designated senior manager, and would include the ICT manager, the librarian, and the guidance counsellor/teacher.

The acceptance of the Internet Safety Kit and the later NetSafe Kit for Schools as models of best practice by the Ministry of Education, and the review of schools' compliance in this regard by ERO, together have had significant impact on schools. Many schools have worked diligently to get the right systems in place in order to meet their legislative obligations under the National Administration Guidelines regarding the provision of a safe physical and emotional learning environment and a school's need to consult with its community. These legal obligations would certainly be true for the virtual classroom as well as the physical one.

Cybersafety education becomes especially crucial with the mobile technology that is going to quickly become commonplace. Some New Zealand schools have made a shift to laptops and wireless technology. Many schools have noticed an upsurge in incidents involving mobile phones, from cyberbullying via text messages to exam cheating. Such concerns have been well-documented recently in the media. When most mobiles also have Internet access and cameras, the situation gets even more complex. By getting the education in place now, concepts of safe and responsible use will transfer more smoothly to the mobile environment where monitoring will be very problematic.

NetSafe training modules for schools
The ISG has developed training modules for those school personnel who have key roles in creating and maintaining a cybersafe learning environment. There are specific modules for trustees, principals, cybersafety managers, ICT managers, guidance counsellors, teachers and librarians. These have been offered by licensed training providers across New Zealand from July 2004. With ERO asking very specific questions about cybersafety policies and procedures as part of a school's regular review, it is important that these key people within a school feel both informed and supported in carrying out their crucial teamwork.

Parents in Cyberspace
More parents are becoming comfortable with communication technologies, but some are lagging far behind their technologically talented children. Yet most parents have important innate guidance to offer their children in all situations, including cyberspace. This is especially true in the case of some of the moral and ethical dilemmas children may encounter.

By offering education and support to parents, schools can gain powerful allies in cybersafety education. If Internet practices in the home mirror what is being done at school, the job of creating a cybersafe learning environment becomes much easier. The goal is to create a cybersafety culture – one that children and young people themselves support – and this can be more easily achieved with a strong alliance between parents and schools (such a culture can also be reinforced by law enforcement, businesses, the media and other community organisations).

Children and young people have numerous options for accessing the Internet, such as cafés and libraries. In fact, a New Zealand survey showed 43 per cent accessing the Internet at a friend's house (Internet Safety Group, 2002). This means that in cyberspace, the parental supervisory role is somewhat restricted. Parents have a difficult job of controlling what material their children can, and cannot, access online. So the parental role evolves to become less one of censorship and more one of preparing children for what they may encounter so they react wisely.

It is admittedly a difficult balance between informing children about what they *may* encounter and giving them information that is difficult for them to handle (that is, regarding child pornography). That is why the advice that younger children should turn off the screen and get help if they encounter something that frightens or upsets them is helpful. The ISG has developed a safety button that can easily be downloaded and installed on computers used by younger children and makes covering the screen and getting assistance a simple action.

The British survey that found one in four children were being bullied or harassed via these technologies also found that 29 per cent of the victims were not telling anyone about the abuse (NCH, 2002). Though more research needs to be done, three factors may be involved here: first, the trusted adults that children and young people might turn to normally for support may be viewed as not being very 'tech-savvy'; second, young people may believe the myth of anonymity on the Internet and think it simply is not possible to catch perpetrators in cyberspace; and third, and perhaps most importantly, children may be afraid that parents will react to a problem by taking away the technology. Thus, education with parents needs to include – as with many other difficult issues such as drinking and driving – the importance of keeping positive communication open and not initially appearing to react too strongly, even when very worried.

Educators can disempower parents if their advice is too simplistic. Putting computers in the family room is strongly recommended, but, at the same time, parents need to be aware of the capability of their child's mobile phone. Installing filtering or using a filtering service can help with exposure to inappropriate material, but it can be relatively easy to circumvent and deals only with the issue of inappropriate or offensive content and not some of the other risks. Telling parents how to check the history of sites visited also can help, but many children are adept enough to manipulate that history any way they want. Parents do not need to have the same skills as their children, but they do need to be involved in their children's use of the Internet and be aware of how much time is spent online and what children are doing.

Schools can support parents by holding parent information evenings. At such events, parents can learn about cybersafety from a guest speaker and can talk with fellow parents and share concerns and strategies. Putting regular cybersafety tips in the school newsletter or on its website can also help. Schools miss an important opportunity if they leave parents out of the cybersafety equation.

Cybersafety in the Classroom

As schools work to get comprehensive systems in place to maintain their cybersafe learning environment, they must also be addressing the issue of cybersafety education.

In a physical classroom, there are already free resources from the ISG that can be distributed and displayed, such as an electronic crime brochure (produced with the New Zealand Police), a parent brochure (produced with the Police Youth Education Service), a brochure and poster on text bullying (produced with Vodafone NZ), and an Online Safety Rule card and posters for primary- and secondary-level students. The amount of such supplementary material available with be increasing rapidly in the next few years. The NetSafe website will have a whole section devoted to classroom learning with links to international programmes which also offer excellent resources.

Cybersafety education that fosters safe and responsible use must begin at the pre-school level. Yet many in this age group (and early primary) cannot read. The ISG has a major project underway to create online animated episodes to teach cybersafety using the NetSafe dolphin character, Hector Protector, and his sea creature friends. Lesson plans and other materials will be available from the website to use in conjunction with the purchased CDs of episodes. Kids relate so well to animation, which offers many advantages, including being able to reach the non-readers and to record the soundtrack in multiple languages. Episodic plot development allows for unlimited expansion and thoughtful dialogue can help children move beyond simplistic 'black and white' notions of cybersafety to better understand the complexity of some of the situations they may find themselves in.

Moving beyond safety messages that stress 'don't, don't, don't' to concentrate instead on building critical thinking skills will be very important with the coming 'morph' to mobile technology and mostly unmonitored Internet access. The ISG's online safety rules stress needing *to stop, think, ask* oneself key questions and *remember* important facts about cyberspace before acting. These rules also include considering how one's behaviour may affect others.

The Police Youth Education Service (YES) is now incorporating cybersafety into both their Keeping Ourselves Safe and Kia Kaha (anti-bullying) programmes in classrooms across the country. The ISG has trained YES officers in every region of the country to offer this important cybersafety education through their grassroots network of 120-plus officers. The officers work with students and parents and can support schools in their use of the NetSafe Kit for Schools.

As you move into the secondary level, cybersafety (and the social impact of communication technologies) is appropriate in a number of curriculum strands, such as health education, social sciences, and computer science. One- to two-week modules are being developed to facilitate that delivery. In tertiary education, cybersafety and social impact topics are appropriate in law, commerce, psychology, sociology, computer science, teacher education and more.

At a minimum, a review of the school's or institution's policy regarding acceptable use at the start of any course using the Internet should be instituted. This is akin to reviewing lab safety rules at the beginning of a science course. Display of cyber-safety posters near any computer access is also recommended. Use agreements are educational documents and every student and staff member should sign one before they get log-on details (preferably at enrolment/employment), and a sample could also be posted near the computers. A reminder about the policy could be used as a screen-saver or as a kind of 'terms and conditions' document that a user must agree to each time he or she begins a session.

For students learning online, similar practices can be instituted. The policy document can be online all the time for easy referral and a link to NetSafe and/or other safety websites could be added to any educational website as a public service.

There are students who will be more reliant on these technologies than others. Members of the deaf community can rely heavily on text messaging and e-mail. Physically and intellectually disabled students can find chat environments wonderfully liberating because they are not judged so readily by their disabilities. Rural students may find the technologies provide a powerful antidote to a feeling of social isolation. In the experience of the ISG, those living is small rural communities can sometimes misunderstand the Internet and believe they are actually safer on the Internet than their urban counterparts are. Gay young people may be able to 'come out' on the Internet long before they can discuss their sexual orientation with family members or peers. It will be important to tailor specific cybersafety information to any group of students who are more involved with, or dependent on, communication technologies.

The Education Frontier

Introducing children, young people and adults to the wonders of the Internet engenders a very real responsibility to prepare these individuals to use the technologies wisely. Teachers utilising e-learning are at the forefront and have demonstrated brilliantly how to incorporate communication technologies into classroom learning as well as how to move the whole learning experience online. Using the same creative initiative that propelled e-learning to the cutting edge of education, online classrooms can now incorporate cybersafety and nurture not just capable e-learners but aware and responsible e-citizens.

References

Berson, I.R. (2002). *Grooming cybervictims: The psychosocial effects of online exploitation for youth.* Technical Report 172, Department of Computer Science, University of Auckland, 9-19. Available at: <http://www.cs.auckland.ac.nz/~john/NetSafe/I.Berson.pdf>

Brown, D. (2001). Developing strategies for collecting and presenting grooming evidence in a high tech world. *National Center for Prosecution of Child Abuse Update, 14*(11). Available at: <http://www.ndaa-apri.org/publications/newsletters/update_volume_14_number_11_2001.html>.

China Daily (2002). As reported on VNUNET at: <http://vnunet.com/News/1134416>.

Ferris, J.R. *Internet addiction disorder: Causes, symptoms, and consequences*. Available at: <http://www.chem.vt.edu/chem-dept/dessy/honors/papers/ferris.html>

Finkelhor, D., Mitchell, K.J. & Wolak, J. (2000). *Online victimisation: A report on the nation's youth*. Alexandria, VA: National Centre for Missing and Exploited Children.

Greenfield, D.N. (1999) *Virtual Addiction: Sometimes New Technology Can Create New Problems*. Available at: <http://www.virtual-addiction.com>

Griffiths, M. *Internet Gambling: Preliminary Results of the First U.K. Prevalence Study*. Available at: <http://www.camh.net/egambling/issue5/research/griffiths_article.html>

Harvey, D.J. (2002). *Internet safety: Young people and the law in an on-line world*. Technical Report 172, Department of Computer Science, University of Auckland, 22–30. Available at: <http://www.cs.auckland.ac.nz/~john/NetSafe/Harvey.pdf>

Internet Safety Group (2001). *Girls on the net: The survey of adolescent girls' use of the Internet in New Zealand*. Available at: <http:/www.netsafe.org.nz/Doc_Library/girlsonthenet.pdf>

Internet Safety Group (2002). *The net generation: Internet safety issues for young New Zealanders*. Available at: <http:/www.netsafe.org.nz/Doc_Library/net_gen_report.pdf>

Joinson, A. (1998). Causes and implications of disinhibited behaviour on the net. In Gackenbach, J. (1998). *Psychology and the Internet: Intrapersonal, interpersonal, and transpersonal implications*. California: Academic Press.

Manch, K. & Wilson, D. (2003). *Objectionable material on the Internet: Development in enforcement*. Available at: <http:/www.netsafe.org.nz/Doc_Library/netsafepapers_manchwilson_objectionable.pdf>

McCarthy, J. (2002). *'Stranger danger' in the home – Managing the risks to children posed by the Internet*. Technical Report 172, Department of Computer Science, University of Auckland, 100–105. Available at: <http://www.cs.auckland.ac.nz/~john/NetSafe/McCarthy.pdf>

NCH (2002). *One in four children are the victims of 'on-line bullying' says children's charity*. Press release. Available at: <http://www.nch.org.uk/news/news3.asp?ReleaseID=125>

O'Connell, Rachel. (2003a). *Fixed to mobile Internet industry: The morphing of criminal activity on-line*. Available at: <http://www.netsafe.org.nz/Doc_Library/netsafepapers_racheloconnell_mobile.pdf>

O'Connell, Rachel. (2003b). *A Typology of child cybersexploitation and online grooming practices*. Preston: Cyberspace Research Unit, University of Central Lancashire.

Reid, E. (1998). The self and the Internet: Variations on the illusion of one self. In Gackenbach, J. (1998). *Psychology and the Internet: Intrapersonal, interpersonal, and transpersonal implications*. California: Academic Press.

Sproull, L. & Kiesler, S. (1986). Reducing social context cues: Electronic mail in organizational communication. *Management Science, 32,* 1492–512.

Wallace, N. (2003). *Net helps children start sex attacks*. Available at: <http://www.smh.com.au/articles/2003/11/25/1069522606196.html>

Combating Online Plagiarism: Closing the Lid on Pandora's Box

12

Shannon Curran and Merrin Crooks-Simpson

This chapter provides a guide of web-based resources on online plagiarism. A brief discussion is included regarding what constitutes both offline and online forms of plagiarism, why it is wrong, and the extensiveness of the issue. We look at various ways of dealing with online plagiarism in secondary schools, and then examine several methods of preventing online plagiarism, such as plagiarism education, relevant lesson plans and tutorials, correct referencing and citation styles and assessment design.

The aim of this resource guide is to assist secondary teachers to understand the issues surrounding online plagiarism, including how they can detect it, and how to inform their students of ways to conduct their work honestly. A number of annotated websites are also included for use by secondary students to aid their own writing practice.

A high number of the detection and prevention strategies outlined in this guide are suitable for teaching students about both traditional and online methods of plagiarism. Therefore, we have tended to use the term 'plagiarism' in these instances to refer to both methods.

What is Plagiarism?

As an example of an educational definition, plagiarism is defined in the *Northwestern Michigan College Catalogue and Student Handbook* (NMC Library, 2002) as: 'offering as one's own work, the words, ideas or arguments of another person, without appropriate attribution by quotation, reference or footnote. Plagiarism occurs both when the words of another are reproduced without acknowledgement, and when the ideas or arguments are paraphrased in such a way as to lead the reader to believe that they originated with the writer' (para. 1).

Plagiarism covers a broad range of acts, from directly copying passages of unpublished or published work by another into one's own work (which is what often springs to mind when considering plagiarism), the presenting of another's work completely as your own, or paraphrasing another's work without correct citation, to the misappropriation of someone else's ideas. Knowing how and when to use citations is not a straightforward process: for instance, what happens when a peer proofreading your work provides you with a new insightful interpretation? Do you claim the idea as your own, or recognise the other person's input as an endnote or acknowledgement? Conversely, when is information 'common knowledge' and therefore does not require citation? Students require a 'toolbox' of relevant skills to be able to correctly cite the ideas of others and to know when it is necessary to do so.

Why is Plagiarism Wrong?

Plagiarism is embedded within legal and ethical contexts. Plagiarism can be considered morally wrong in the same way as physically stealing food, money or other items is viewed by society (Petress, 2003). As DeVoss and Rosati (2002) note, this is strongly related to the laws surrounding intellectual property rights. In other words, intellectual property rights make acts of plagiarism illegal and the plagiarist is committing an act of theft (Gibelman *et al.,* 1999).

Although not directly obvious, plagiarism also affects the educational experiences of all students: for example, in the worst-case scenario, comparisons between the work of undetected plagiarists and other students can lead to plagiarised work being regarded as of a higher standard (Ryan, 1998, cited in Underwood & Szabo, 2003).

What is   Plagiarism and How Does it Differ From   Plagiarism?

What is meant by online plagiarism? Online plagiarism is primarily the same act as offline plagiarism; however, in the case of online plagiarism, the material is sourced electronically – possibly from the Internet, which itself introduces new elements of complexity (Underwood & Szabo, 2003). This opens a virtual 'Pandora's box' in terms of both deliberate and accidental acts of plagiarism by students attempting to use the ideas of others within their own work.

The information revolution and consequent explosion in the use of computer-mediated tools is changing students' approaches to gathering and using information for educational purposes. In terms of deliberately plagiarising, the simple 'cut and paste' that is used within document manipulation is easily transferable to the online environment and can result in online content being effortlessly manipulated into new creations for students' use (DeVoss & Rosati, 2002; Galus, 2002). Additionally, Underwood and Szabo (2003) suggest that by using the Web, students can be more inventive in their techniques for plagiarising the ideas of others. Within online environments, it is possible to simply observe interactions or more directly ask for assistance from discussion groups and accordingly take ideas for one's own benefit without ever acknowledging its origins (Underwood & Szabo, 2003).

It is important to remember that some acts of online plagiarism can occur accidentally, and this is where plagiarism education strategies for students become so essential. The information displayed on the Internet is more complex with such features as graphics, audio and multiple links making authorship hard to ascertain let alone cite correctly (DeVoss & Rosati, 2002). Furthermore, what defines plagiarism in various contexts is determined by specific cultural, historical and work contexts, which can make the navigation of this issue difficult (Gajadhar & Brooke, 2001; Price, 2002): for example, in some non-Western cultures, the most revered material that is considered 'original' is material that was historically created, and new thoughts are considered less valuable (Fox, 1994, cited in Price, 2002).

Underwood and Szabo (2003) propose that educators themselves need to consider the integrity of their own work, as what constitutes plagiarism is often hard to grasp for both students and teachers. For example, have you ever 'borrowed' or adapted a lesson plan without giving appropriate attribution to its origins? Or emailed that funny cartoon picture to share with friends without considering the copyright laws that may prevent such activity? Few people (including the authors of this chapter) could ever claim to be completely 'squeaky clean' of plagiarism in all aspects of their lives.

How Extensive is Online Plagiarism?

The availability of billions of pages of information that is accessible literally at the click of a mouse, that is up to date, easily downloaded, and potentially difficult for educators to detect, means plagiarism is becoming more tempting for students and consequently more of a problematic issue for educators (Gajadhar & Brooke, 2001). Not only can information on practically any topic be searched for and downloaded, but there are a huge array of websites, known as 'paper mills' or 'paper dealers' claiming to cater specifically to students 'educational needs', and providing them with either customised or on-file essay papers, many free of charge.

Relevant studies indicate that plagiarism is a significant problem in today's educational institutions, and that the magnitude of the problem has increased in recent years (Brown & Howell, 2001; Gibelman *et al.,* 1999). This can be partially attributed to the fact that some students seem to view almost anything that they discover on the Internet as authorless 'general knowledge' that does not require citation. Some students assume that they can download music and read articles for free on the Internet; this attitude has extended into their academic writing, where they see it as acceptable to incorporate passages of text into their written assignments without acknowledgment of where they came from (Young, 2001). Additionally, educational reforms have led to a more competitive education system for students, and higher student-to-teacher class ratios make the surveillance of student practices more difficult to undertake (Underwood & Szabo, 2003).

How often does online plagiarism occur? McCabe's (2001) study of 2294 students attending US schools at the junior-high level (approximately aged seventeen) found that 52 per cent admitted taking a few sentences from online sources without correctly attributing authorship and 16 per cent admitted using essays largely sourced from paper mills or websites. In contrast, at the tertiary level, Scanlon & Neumann (2002) found that 24.5 per cent of their 698 participants admitted to online plagiarising sometimes to very frequently. In the same study, 6 per cent of students admitted to sometimes purchasing a paper online, while 2.3 per cent admitting to doing so frequently.

Why does online plagiarism seem to be more prevalent at secondary level? McCabe (2001) answers this by noting that students at secondary level are still learning about the concept of plagiarism and how to cite material correctly. However, on a more worrying note, online content is viewed by secondary students as being easy to obtain and difficult for teachers to detect within plagiarised work (McCabe, 2001).

Unfortunately, most research studies regarding the incidence of online plagiarism are situated within the wider context of academic dishonesty and are usually concerned with the tertiary context (Scanlon & Neumann, 2002). It is also possible that students' belief that online plagiarism is prevalent, based on anecdotal evidence, could prove to be self-fulfilling with students deciding that such acts are acceptable because everyone else does it (Scanlon, 2003). For these reasons we would suggest that the area of online plagiarism needs more research specifically at the secondary level.

So where to from here? This resource guide offers both strategies for dealing with online plagiarism, and strategies for preventing online plagiarism. We hope it can serve its purpose in helping educators to close their own version of Pandora's box in their classrooms.

Strategies for Dealing with Online Plagiarism

Participating in plagiarism discussion groups / online forums

Electronic discussion groups are asynchronous forums that allow geographically dispersed participants to interact, collaborate and discuss areas of interest. Participants can read postings, start new discussions or contribute to existing ones.

For teachers and educators, discussion groups offer two major advantages. First, participants can choose to contribute at convenient times based around their lifestyles and work commitments; and second, they offer the prospect of communication with peers outside the participants' places of work, and therefore, the opportunities to perhaps gain fresh perspectives (Hawkes, 2000).

For students, discussion groups offer the opportunity to develop their own understandings of plagiarism in a relatively anonymous environment. Vonderwell (2003) notes that for students, the relative anonymity of online asynchronous communications removes the possible anxiety that some students feel asking questions in front of their peers, and that the act of writing helps students to clarify and reflect on their ideas.

When constructing this resource, discussion groups were limited to those that specifically dealt with the issue of online plagiarism or were focused on the use of technology for educational purposes in the context of secondary education.

The named discussion groups are moderated (meaning that any submissions go to an editor in charge of the network, who will remove any unsuitable or unrelated posts), are free to join, provide some form of online help and have search facilities for browsing through their archives. New participants can join by filling out online subscription forms.

H-Net Discussion Networks: EDTECH (Educational Technology)
http://www.h-net.org/~edweb/

H-Net is an international organisation formed by teachers and academics from various disciplines that is aimed at exploring the educational potential of the Internet. H-Net provides a number of discussion networks organised into topic areas for students, teachers and university lecturers who may ask questions and discuss relevant issues. The discussion networks offered cover a wide variety of topics within the divisions of humanities and social sciences such as evaluation and recommendations for software, problems and solutions related to the use of educational technology, educational technology conferences and research information.

At the end of this particular web page, there is a useful link to an Internet citation guide. EDTECH is just one of the networks available and is aimed at users of technology in education. EDTECH would be most suitable for secondary teachers with an interest in educational technology.

JISCMail: Plagiarism discussion list
http://www.jiscmail.ac.uk/lists/plagiarism.html
The Joint Information Systems Committee (JISC) funded by the British post-sixteen and higher education funding groups, offers a wide variety of moderated discussion groups organised into various areas of interest, including this one which focuses on plagiarism.

The intended subscribers to this discussion group are from the academic research, university lecturer and secondary school teacher communities. The site aims to promote collaboration and discussion amongst educators. This particular discussion group focuses specifically on issues surrounding plagiarism detection and prevention, and has a multi-discipline focus.

Other issues that have been discussed within this group include: research into plagiarism; teaching appropriate detection and prevention strategies for educators; comparison of detection methods, and the actual incidence of plagiarism.

Using plagiarism detection software
As an educator, if you suspect that part or most of a student's work has been plagiarised from online sources, the most simple detection strategy (and also, arguably, the cheapest) is to use a search engine such as Google (www.google.com) to search for matching strings of words used within the student's work.

In order to combat the growing number of paper mills and paper dealers that are in operation supplying student demand, an industry has sprung into being that develops plagiarism detection software. Despite the various software manufacturers providing online 'endorsements' of their products, several issues remain, and discretion is advised in using them. The usefulness of such software is questionable, since searches are only conducted for material that is available on the Internet or a database of student-submitted work (Royce, 2003). Material plagiarised from other sources, such as radio, television, CD-Roms, books, magazines or newspapers, are not going to be found (Royce, 2003). Research is needed to assess whether the institutional use of plagiarism software effectively discourages students from attempting to plagiarise using online material.

Plagiarism software generally requires that students submit their work in an electronic format. The software will then often remove the formatting features within their work. This being the case, students may be asked to submit their work both electronically and in printed form. This creates a number of issues. First, how do you ensure consistency between the electronic copy and the printed version? (Centre for the Study of Higher Education [CSHE], 2002). Second, it is important to consider the ethical balance between the rights of students to protect their own work and the need to maintain educational integrity. For instance, turnitin.com maintains an ongoing database of student work used for making comparisons (Foster, 2002). With regards to ethics, it is important to gain informed consent from students before storing electronic copies of their work (Foster, 2002).

Plagiarism software (especially freeware) may seem to be appealing, but before using any plagiarism software, it is important to check out its origins. There have been concerns expressed in the media regarding possible links between some detection products and several paper mills (Young, 2002). We advise that wherever possible, use software based on recommendations from your colleagues or available institutional guidelines, and be sure to conduct a broad Internet search for commentary on the software.

The Centre for the Study of Higher Education is an Australian organisation based within the University of Melbourne that offers an excellent overview of several plagiarism software products, including a comparison of features and a discussion on their effectiveness. This is available from <http://www.cshe.unimelb.edu.au/assessinglearning/03/Plag2.html>. Table 1 opposite lists some examples of plagiarism detection software.

Understanding what paper dealers and paper mills are

The general availability and use of the Internet for educational purposes has allowed students access to an internationally accessible virtual library of resources for their academic assignments. Unfortunately, a number of enterprising developers have recognised a commercial niche for providing students with ready-made assignments; hence the growth of 'paper dealers', also known as 'paper mills'. These sites are generally intended for a university-based audience, but there is nothing to stop an 'enterprising' secondary student from using their services (Royce, 2003).

How easy are paper mills to use? The answer is *very*. And there are lots to choose from. All students need to do is provide payment. Some paper mills even offer a service that, for an increased price, will produce an essay on any topic required. As an example, the eighty-five keyword search/subject areas covered by Essay Find range from accounting to zoology and include book reports, global politics, history, sciences, sports issues and Shakespeare (to name a few).

These purchased essays vary in standard. Basically, the more money an individual is willing to spend, the better the quality of the essay (Gibelman *et al.,* 1999). Regarding the merit of such purchases, the authors of this small study comparing the marks of three essays (two bought online and one legitimately created) that were blindly assessed by tertiary educators found that such essays could potentially pass, although with lower results.

Most paper mills attempt to protect themselves from prosecution by including a disclaimer asking students: not to submit these papers as their own, but rather to make use of the citations and reference lists included to gather their own material; and to be careful to reference any contact they may use directly to its original source.

Some examples of online paper mills include:

• A1 Termpapers <http://www.a1-termpaper.com/>

Table 1: Examples of plagiarism-detection software

Eve2 Essay Verification Engine	http://www.canexus.com/eve	Available for a free trial period. Individual or site licences are available. Electronically submitted essays are scanned and compared with advanced Internet searches to look for plagiarised sections that have been sourced online. Limited online support is offered.
Turnitin. com	http://www.turnitin.com/	Available for a free trial period. Individual or site licences are offered. Electronically submitted essays are scanned and compared with advanced Internet searches and with the manufacturer's own database to look for plagiarised sections. Submitted essays are kept to add to the database of comparison material. Excellent online support.
Glatt Plagiarism Services	http://plagiarism.com/	Licensing options are not clearly stated. A tutorial is offered aimed at providing students with instructions on how to paraphrase correctly, and how to cite references appropriately. The detection programme differs considerably in its methodology, with students being required to sit a test, filling in missing words from their work to arrive at a 'plagiarism probability score'. With the final product offered, students are able to evaluate their submitted essays online for plagiarism. No apparent online support options are offered.
WCopyfind 2.2	http://plagiarism.phys.virginia.edu/Wsoftware.html	Freeware (able to be used without a licence). Compares student essays with established files that share large amounts of material. When two files have large amounts of common material, the software generates a report with the shared material underlined. This software does not perform wide Internet searches looking for material that has been plagiarised online. Little online support is available.

- Essay Find <http://www.essay-find.com/>
- Evil House of Cheat <http://www.cheathouse.com/>
- Free Papers! <http://www.freepapers.net/>
- Free-TermPapers.Com <http://www.free-termpapers.com>
- Genius Papers <http://www.geniuspapers.com/>
- Other People's Papers <http://www.oppapers.com/>
- Research Papers Online <http://www.ezwrite.com/>
- School Sucks <http://www.schoolsucks.com/>
- Slackers <http://www.7thsphere.com/slcakers/main.shtml>

Cheating 101: Paper Mills and You
http://www.coastal.edu/library/pubs/papermil.html
'Cheating 101: Paper Mills and You' provides an excellent overview of the issues regarding paper dealers and paper mills for educators, and contains information relevant to secondary teachers.

This site suggests prevention as being the best strategy against plagiarism and offers a broad overview of plagiarism issues within secondary and tertiary environments. Issues covered include the incidence of plagiarism, why students plagiarise, and how to deal with plagiarism, plus links to a good checklist of 'symptoms' to look for that may indicate plagiarism <http://www.coastal.edu/library/plagiarz.htm>

When this site first started keeping a track of paper mills in 1999, it had thirty-five sites listed. At June 2004, it listed over 250 <http://www.coastal.edu/library/mills2.htm>.

This site 'dishes the dirt' on paper mills, including the costs involved and, what students are getting for their money, and lists other online sources that students may potentially plagiarise material from.

The Need for Education about Plagiarism

Educational institutions worldwide are taking the obligatory steps to confront the issue of online plagiarism head-on through the use of plagiarism-detection software and other similar means of revealing academic dishonesty. These systems are proving to be successful in exposing those who claim another's work as their own. However, the process can be time-consuming, expensive and trying for those involved. Additionally, some educators are concerned that the use of plagiarism-detection services within education institutions will ruin any trust that was once present between a student and his or her teacher.

There are definitely two sides to the coin on this matter. Gajadhar and Brooke (2001) suggest that the answer lies in the aim of plagiarism prevention: 'Electronic tracing facilities, honesty declarations and written regulations on the penalties of plagiarism are useful deterrents, but real solutions lie in the nature of the learning experience of our educational institutions. To prevent plagiarism, particularly the online variety, students must be encouraged to take responsibility for their actions, and to consider the moral consequences' (pp. 50–51)

Renard (1999/2000) also supports this viewpoint. She states, 'Catching Internet cheaters is not the best answer. It's a lot like doing an autopsy. No matter how terrific the coroner is at determining how or why a person died, the damage has been done. Bringing the culprit to light won't change that. Preventing the problem is a much better approach … No matter how great we get at detecting student plagiarism, we won't be undoing that educational damage' (para. 21)

Educators need not feel helpless in preventing plagiarism. Research by Brown and Howell (2001) suggests that the provision of guidance and instruction about avoiding plagiarism, which encourages students to take a more serious view of the issues, is likely to have positive effects on students' future behaviour.

Research by Landau *et al.* (2002) also indicates that students can quite easily learn to detect plagiarism and avoid it simply through educators giving examples or feedback on paraphrasing attempts. This process has been proven to have a positive effect on students' knowledge of plagiarism and their ability to avoid it.

It is neither difficult nor time-consuming to effect a change in students' abilities to detect and avoid plagiarism. By adopting a proactive approach to eliminating plagiarism, and by discussing with students what exactly constitutes plagiarism and how one can avoid it by using correct citation methods, those students who are unclear about plagiarism and who may assume that they are succinctly knowledgeable will finally get the greater understanding they need.

There is evidence that suggests that some forms of plagiarism might result from students' inadequate knowledge of proper citation techniques (Froese *et al.,* 1995; Roig, 1997, both cited in Landau, *et al.,* 2002). This fact increases the significant benefits that teaching students about plagiarism and how to cite information sources correctly can bring.

Moreover, instructors who assume that students know how to avoid plagiarism may miss an opportunity to teach the skills needed to avoid the consequences of academic dishonesty. By discussing plagiarism with all students, correct knowledge on how to cite references can be passed on, ways for students to see the value in the writing process and to take pride in their own skills and knowledge can be shown, and students can be encouraged to take responsibility for their actions and to consider the moral consequences of plagiarism.

Educating students

Educators should never assume that students understand the concept of plagiarism (DeVoss & Rosati, 2002; Price, 2002). Research has shown that many plagiarisers who are caught claim that they did not know that they were 'cheating' or that they didn't think it was such a serious issue. As Petress (2003) states, 'Such statements suggest that too few people know exactly what plagiarism is; they are unaware of rules against plagiarism; and/or they have learned through benign neglect from teachers, school administrators, school boards, and parents that plagiarism is not a big deal' (p. 624).

It is imperative that educators teach students skills in paraphrasing, proper referencing and citation, for both written and online sources. Students who have yet to master the finer arts of paraphrasing may submit work that contains 'patchwriting'. Patchwriting occurs when students leave the referenced material fundamentally intact by only making small changes – such as grammar changes within a sentence or using words of similar meaning to the original (Price, 2002). This is a form of plagiarism and can be curtailed through providing sufficient paraphrasing practice.

It is important that students be taught at least one common method of citation (e.g. APA style) and be required to use it habitually in their work. As Galus (2002) suggests, 'Teachers should first explain that whenever students use information and ideas from another source they must always give the author credit for the material. Failure to give credit to the original author, whether for paraphrasing or using a direct quote, is plagiarism' (p. 37).

Educators need to follow this up with a sound explanation of the consequences of plagiarism within their educational institutions. Petress (2003) suggests the need for an overall institutional policy and dialogue between school and parents regarding the issues that surround online plagiarism.

Students should be given the opportunity to discuss the notions of authorship (DeVoss & Rosati, 2002; Price, 2002). Teachers should also encourage students that their own works, including their own ideas, are important both to themselves and to others (DeVoss & Rosati, 2002).

Plagiarism is not just an 'academic' issue, there are many real-life examples that can be used to form the basis of classroom discussion regarding authorship and ethics. An example of this is the discovery of the *New York Times* reporter Jayson Blair's liberal use of other people's material within several articles and the subsequent scandal that emerged during 2003 (see Boehlert, 2003, for more details on this incident). Additionally, DeVoss & Rosati (2002) give the example of the ongoing issues regarding the availability of music online, including the shutdown of free-provider Napster due to legal action.

Furthermore, students should be encouraged to be critical of material that is available from online sources. It is important for students to acknowledge that not all online material is necessarily well-founded based on a number of factors. DeVoss and Rosati (2002) suggest students should examine three factors: the relevancy, appearance and authorship of online material. The relevancy of online material can be questionable, depending on how often the website is updated. Students need to learn to question whether the graphics on the site merely make it look pretty, or whether they add some value to the contents. In addition, it is also important for them to consider who set up the website, and if they have any agenda that may make the material impartial. Educators could discuss techniques of evaluating online content with their class and prepare an evaluation checklist as a class activity. An example of a pre-prepared checklist is available from the University of Otago website at <http://www.library.otago.ac.nz/pdf/wc_checklist_evaluation.pdf>.

Assessment design

We strongly support the position that good assessment design can reduce the potential for students to plagiarise (Galus, 2002; Gibelman *et al.,* 1999; McMurtry, 2001; Petress, 2003). Research has shown that the largest percentage of academic grades is based on written assessments (Gibelman *et al.,* 1999). This evidence is disruptive to the cause of plagiarism prevention given that using a *variety* of assessment techniques has been shown to reduce the incidence of plagiarism: for example, assessments in the form of oral presentations require students to be very familiar with their chosen subject (McMurtry, 2001). It is the responsibility of teachers to consider how the design of their assessments can be developed to reduce the possibility of their students plagiarising (Gibelman *et al.,* 1999; Petress, 2003).

When the style of assessment *does* involve written work, however – for example an essay or a report – it is recommended that teachers be 'creative' with topic choice and take the time to not only provide students with themes that are intrinsically interesting but that also require the synthesis of ideas (McMurtry, 2001). This is supported by Gibelman *et al.* (1999) who note, 'if assignments are oriented to integrating diverse educational experiences, the availability of cheating options is substantially reduced' (p. 374). By limiting the topics a student may write about, and by giving specific instructions, the more difficult it will be for anyone considering plagiarising to find relevant material that fits the assignment (McMurtry, 2001). In order for this to be most effective, educators need to re-examine the assignment topics they set each year that they teach a module in order to further avoid the simplicity of copy-and-paste plagiarism by students (Gibelman *et al., 1999).*

Supplementary ideas to aid in the prevention of plagiarism in written assignments include: students being asked to provide printouts of original online sources (Galus, 2002; Royce, 2003); requiring students to submit outlines or working drafts of their assignments throughout the writing process (Gibelman *et al.,* 1999); having teachers carry out their own Internet search on a specific topic and becoming familiar with what is available online (McMurtry, 2001); performing spot-checks of citations students have used; and having students submit their assignments electronically in order to keep a database of students' work (McMutry, 2001). The latter is especially useful for teachers who for some reason have been unable to change an assignment topic from previous years, and who suspect plagiarism. However, as previously mentioned, there are ethical issues that need consideration when storing student work, and student consent should be obtained (Foster, 2002). As Galus (2002) suggests, teachers need to be actively involved in *all* aspects of research, not just the setting of a topic.

It is important to acknowledge that even if possible plagiarism is detected, proving or disproving that students have committed plagiarism is not a black-and-white issue. As Royce (2003) notes, teachers need to have good research skills and to be familiar with what material is available on a given assignment topic in order to be able to follow up on material that intuitively seems out of place in student work. Educators need to find an *exact* match to the original source of the material, and then show that correct attribution has not been given, or that the one given is wrong (Royce, 2003). For this reason, we stress the importance of online plagiarism *prevention* as the best cure to this new issue in academic circles.

Plagiarism lesson plans and tutorial websites

The following is a list of annotated websites that offer lesson plans for teachers on educating about plagiarism, and websites to refer to for correct citation methods.

ICT Learning Experience: A Plague on Plagiarism
http://www.tki.org.nz/r/ict/ictpd/plague_on_plagiarism_e.php
This resource is provided by the popular New Zealand-based Te Kete Ipurangi (TKI) website. It provides a flexible lesson plan aimed at teaching students to take notes in order to avoid plagiarising other people's work. Students are educated about copyright and plagiarism issues, and are taught how to improve their research techniques. Although aimed at lower secondary-school students, the main principles

can be applied at all levels of education. Useful links are provided to PDF documents that assist in the teaching of this lesson.

The evaluation section documents the effectiveness of this lesson plan for the classes who have used it. Almost all of the Year Nine to Ten students surveyed after receiving the lesson felt that they had achieved better results in note-taking, that they liked having such a process to follow, and that what they had learned would be very useful in the future.

RIT Library: Copyright and Plagiarism Tutorials
http://wally.rit.edu/instruction/dl/cptutorial/
This site offers online tutorials and information suitable for senior students and secondary teachers. The preliminary exercises provided for both groups not only ask the participant to decide if the given example is plagiarised but also *why* it is plagiarised.

The site clearly outlines the hardware and software requirements for using its resources. Its one trap is that for those who are non-technical, the online PowerPoint presentation supplied for students can prove time-consuming to run successfully. The PowerPoint presentation provided for academic staff includes a good outline of the main issues surrounding plagiarism, including: definition of plagiarism; discussion of the link between plagiarism and copyright; tips for staff on how to detect plagiarism; and examples of the Rochester Institute of Technology's plagiarism policy. The site also includes a paraphrasing exercise, which could be adapted for use with senior secondary classes.

University of Washington: Avoiding and Detecting Web Plagiarism
http://depts.washington.edu/trio/train/howto/avoid/index.shtml
This site offers resources suitable for secondary and tertiary students and their educators. It focuses on how to avoid committing plagiarism, particularly within the context of designing and building websites. It also gives advice for students on how to cite material correctly on their websites, and offers a good checklist for them to follow, including best-use practices for audio and visual resources. For educators, there are useful detection tips and a link to other plagiarism resources. Most of the material offered is available for printing or downloading in both PDF and Microsoft Word formats.

CyberSmart: Whose is it Anyway?
http://www.cybersmartcurriculum.org/lesson_plans/45_08.asp
The lesson plans on this site are available free of charge to educators as long as they obey copyright by not altering the content, give appropriate credit to the site and do not distribute the material for financial gain.

The specific lesson plans are designed for a younger audience; however, the issues and material it raises could be adapted to work with older secondary students in any curriculum area. The lesson objectives include: promoting a discussion of what constitutes plagiarism; issues relating to the availability of material on the Internet; and when it is okay to use such material. The plans may be downloaded in PDF format.

Daily Lesson Plan: Cite Your Sites

http://www.nytimes.com/learning/teachers/lessons/20010629friday.html

This website offers a 45-minute lesson plan that alerts students to how the Internet has increased the prevalence of plagiarism, and then gives them the opportunity to practise properly citing sources in their own research-based writing. The lesson is aimed at students of secondary school age.

The lesson guidelines are very precise and easy to understand. Specific learning objectives are included, along with a detailed list of resources/materials that will be required, further questions for discussion, an evaluation and assessment section, some extension activities, a vocabulary extension list, suggestions on how the lesson can be connected to other disciplines, and links to other websites with complimentary material on them.

This is an excellent, easy-to-follow lesson plan for teachers of all subject areas to utilise in the lower secondary school classroom.

Lessons to Teach Students About Plagiarism

http://1rs.ed.uiuc.edu/wp/copyright-2002/lessonplanfaqs.html

This website claims that there are other ways to teach students about plagiarism besides class discussions. It offers a comprehensive list of links to some of the many creative and unique ideas the web offers in regards to plagiarism education.

The links are divided into sub-sections based on school age: high-school lessons, middle-school lessons, elementary school lessons, and K-12 lessons, followed by a list of links to relevant websites on educating students about plagiarism.

There are three links under the high school lessons section. The first includes a multi-choice quiz on whether certain passages of text have been plagiarised or not when compared to the original; the second provides a one-hour lesson plan for Year Nine to Twelve students that aims to teach students the definition of plagiarism and how to avoid plagiarism in their own work. The third link is no longer live.

The further website links are useful, but you would require a fair amount of time to search them all and extract what was relevant to your specific needs.

The Lemonade Tutorials: Plagiarism

http://www.coedu.usf.edu/~dorn/Tutorials/lemonade.htm

Be warned, the author of this site Sherman Dorn, has a mischievous sense of humour. A tertiary academic within the education field, Dorn wrote this site's content and gathered the included resources; however, it would also be useful to senior secondary teachers. This site is not an actual tutorial or lesson plan as such, but is worth including as it gives a good indication of the reasons students might use to justify their plagiarising and examples of possible responses, in addition to a good list of resources for educators.

Correct referencing / citation style websites

University of Otago Library: Citation Styles

www.library.otago.ac.nz/resources/virtref.html#cite

This is the page supplied by the University of Otago for students who have citation queries. Various links are supplied to address different citation styles. For example, there is a link to a page dedicated to geography students on how to cite material within the Geography Department, as well as a similar page for the Law Faculty. With a bit of searching, you will discover there are links to books that address citation styles for other schools and departments within the University of Otago. Library shelf references are also provided for books written about certain referencing styles.

The most useful aspect of the site is the links to PDF documents providing detailed examples of how to reference specific text information sources, such as books, journal articles and newspaper articles, and the PDF document on how to reference electronic resources. However, it took a while to find these links.

This page would be very useful to both students and educators, but it would help to know exactly what you are looking for before you begin using it. For secondary students, access to this site would provide good practice for those who choose to pursue future tertiary study.

Turnitin.com: Citation

www.turnitin.com/research_site/e_citation.html

Part of the renowned Turnitin.com website, the page has been developed for the use of both teachers and students, and offers a definition of what citation is and then outlines the reasons when and why utilised sources should be cited.

In-depth but straightforward instructions are provided on: how to cite sources in the body of a paper; how to incorporate quotes correctly; and how to create a bibliography list. The site also includes an extensive page on citation styles, as well as links to directions on how to use footnotes and endnotes. This site is easy to follow and thorough. We would highly recommend it to both teachers and students of all levels.

APA Style.org

www.apastyle.org/

APA Style is a website dedicated totally to the common citation style established by the American Psychological Association, known as 'APA'. The site defines APA style and outlines its origins.

It would be extremely useful to educators as it not only offers correct APA citation methods but it also has links to pages that discuss the ethical issues surrounding plagiarism, and information on how to remove biases such as on disabilities, sexuality, race and ethnicity in language.

The electronic resources link would be useful to students because it provides instructions on the correct way to cite material obtained online. The electronic media spelling guide link is also useful: it shows the correct way to spell some common electronic terms that can be quite confusing for both teachers and students, such as e-mail, URL, and PDF.

Internet Citation Guides
http://library.wisc.edu/libraries/Memorial/citing.htm

This website is like a miniature database of links to other websites about how to site electronic sources. Included in the main headings are APA (American Psychological Association) – eight links; CBE (Council of Biology Editors) – three links; Chicago – five links; MLA (Modern Language Association) – seven links; Turabian – four links, and Other Styles and Resources (including APSA and NLM) – seven links.

Not all these sites include formats or examples of citations, which could prove to be problematic for those wanting to use the sites to clarify citation styles. Although these sites do not specifically provide instructions on how to cite material that is not from electronic sources, many of them do include links to further sites that do.

It would be a matter of trial and error to find useful links from this page, but if you have the time, then it could be worthwhile because the majority of the links would be very useful to both students and teachers.

Basic CGOS Style
http://www.columbia.edu/cu/cup/cgos/idx_basic.html

This is a website that reviews a book called *The Columbia Guide to Online Style* by Janice R. Walker and Todd Taylor (New York: Columbia University Press, 1998). Included in the site are instructions and examples of how to reference electronic sources correctly, using the elements of citation for both a humanities style (that is, MLA and Chicago) and a scientific style (APA and CBE). Instructions and examples are under two main headings: documenting sources in the text; and preparing the bibliographic material.

The layout and presentation of the sight is very bland, and students could lose interest if they were to use this site as part of a lesson, but the information within the site on referencing electronic resources is nevertheless useful. This site could be used by both teachers and students.

Slate Citation Machine
http://www.landmark-project.com/citation_machine/index.php

This is an extremely useful website for citing in MLA or APA style. A list of source types is displayed on the left side of the page (e.g. web page, newspaper article, journal article, interview, CD-Rom encyclopaedia article). When you click on one of these, a list of relevant fields specific to the source type will appear for you to fill out – for example, when you click on the 'interview' source, the fields interviewee's last name, interviewee's first name, type of interview (personal interview, phone interview, e-mail interview, video conference interview), and date of interview (dd mm. yyyy) will appear. Once the fields have been filled in, one click on the 'process' button will generate two references from the information you entered, one in APA and one in MLA style.

There are no specific instructions about how to go about citing correctly, or any information on why citing is important, but the site is very easy to use, and guarantees correct citation of various information sources. This is an excellent website for use by both students and teachers.

A Guide for Writing Research Papers
http://webster.commnet.edu/apa/index.htm
A guide to citing sources in research, this site is directed at tertiary students (although it would also be useful for educators). It is presented in a question-and-answer format.

An introduction outlines the purpose of this resource and offers further links to information not contained within the page, such as links to sites on citation styles other than APA. Links to information on correct grammar and writing could also be useful to both students and educators.

Students' questions are placed under three main headings: students' questions about manuscript preparation; students' questions about references; and students' questions about parenthetical citations. The references section is the most useful, with common questions students ask being answered in uncomplicated ways with good examples provided.

The only disadvantage is that there appears to be no apparent way students can ask further questions. Information is limited to that which is currently on the website.

Plagiarism one-stop shops
What does 'one-stop shop' mean? For the purposes of this resource guide, a 'one-stop shop' is an already established resource guide available on the Internet. Most provide links to topical areas under the umbrella of online plagiarism – such as software sites, paper dealers, prevention and detection strategies, lesson plans, policy guides and information regarding the incidence of plagiarism, and some also provide original content. This selection of sites is suitable for secondary school teachers and university lecturers who want the 'fast-food' version of how to deal with online plagiarism. Some of the sites also provide tips and advice that would be of use to both students and their parents.

Plagiarism Stoppers: A Teacher's Guide
http://www.ncusd203.org/central/html/where/plagiarism_stoppers.html
Put together and maintained by a secondary school librarian, the site informs educators in general of the issues surrounding copyright, citation and thus the avoidance of plagiarism.

The detection section covers the basics, from using an online engine to search for suspected plagiarised passages to offering links to more comprehensive sites that include checklists of symptoms for possible online plagiarism. There is also a good number of links to detection software (both the free and fee-paying types), paper mills, and links to other 'one-stop shops'.

The prevention section covers material ranging from the provision of guides suitable for students on how to cite material correctly to sites that give teachers tips and advice to use within their own teaching practice.

Safety 'Net @2Learn.ca: On Plagiarism

http://www.2learn.ca/mapset/SafetyNet/plagiarism/plagiarism.html

This resource guide on plagiarism was designed *for* teachers *by* teachers. The site includes a good discussion of how the introduction of technology into classrooms has both provided educational benefits and some downsides, one of which is increased opportunities for students to plagiarise from online material.

Links are provided to a wide variety of handouts that can be adapted by secondary teachers for use within their own classrooms. There is also a good section on examples of plagiarism and acceptable-use policies at both the secondary and university level for educators involved in developing institutional policy development. The guide develops and explains a four-step process for preventing and detecting plagiarism based on: inform, detect, respond and contract.

Finally, this site provides an excellent list of resources for teachers and their students – for example, lesson plans for teachers, and appropriate use and citation strategies for students.

Plagiarized.com: The Instructors Guide to Internet Plagiarism

http://plagiarized.com/index.shtml

This original site aims to educate teachers, university lecturers and parents about online plagiarism. It offers training to educators, providing a checklist of symptoms that may indicate plagiarism in their students' work, and a sample of essays obtained from paper mills with an analysis of what would indicate that they have been plagiarised.

For students, there is a good list of how to obtain research resources legitimately, using their library's interloan service and electronic databases, how to post to appropriate news groups to ask for possible reference sources, and how to ask their teachers and educators for pointers.

The site includes an article by Greg van Belle (n.d.) – How Cheating Helps Drive Better Education, which gives a good outline of best-practices for educators to limit the occurrence of online plagiarism. This is particularly relevant to both secondary school teachers and university lecturers who may teach the same course repeatedly. Suggestions include: giving original assignment questions and readings; offering other forms of assessment such as creating posters, giving verbal class presentations; creating process-based assessments; keeping a file of students' work (even just the front page of essays) in order to monitor their progress and writing style during the course; and being proactive about plagiarism in terms of policy and procedure. The site also suggests that a case of plagiarism should also be seen as an opportunity for educators to reflect on their own teaching practice.

FAQs for Educators on Intellectual Property, Copyright, and Plagiarism
http://lrs.ed.uiuc.edu/wp/copyright-2002/indexfaq.html
An Internet version of a white paper written as part of a university student group project in 1999 and revised in 2002, this site covers a broad range of issues including: copyright, plagiarism, and intellectual property.

This site is in a question-and-answer format, and provides links to other online resources. The links are particularly useful: some, such as articles discussing real-life debates concerning incidences of plagiarism, could be adapted for use within lesson plans and examples of student contracts as a method of prevention (e.g. there is a document that clearly outlines what constitutes plagiarism within institutional guidelines which students are expected to sign as an agreement of their understanding).

The downside of this site is that because of the format, it can take a reasonable amount of browsing to find specific material (its lesson-plan section has been described above under 'Lesson plans').

Paradise Valley Community College: Resources for Plagiarism Information
http://www.pvc.maricopa.edu/library/plagiarism/
Created and maintained by the library staff at Paradise Valley Community College in Arizona, the site is a collection of links to Internet sites that was created for the use of their teaching staff. As such, the site would be suitable for secondary teachers and university lecturers. Please note that some of the links provided, such as to the online database search results, do not work for viewers who are off-campus, but this only rules out a couple of the links provided.

The site includes a definition of what constitutes plagiarism within Paradise Valley Community College, coverage of prevention and detection strategies, lesson plans, plagiarism-detection software, institutional policies, paper mills and news articles in the news about online plagiarism.

Reference and Instructional Services: Preventing Plagiarism
http://www.library.ucla.edu/yrl/referenc/plagiarism.htm
This website is part of the UCLA Library homepage. It provides numerous links to websites, articles and books related to the issues surrounding plagiarism. The main headings include: informative websites on plagiarism; strategies for preventing plagiarism; teaching students about plagiarism; how to detect plagiarism (including full text articles on this matter); term paper mills; UCLA resources (for both teacher and student); recent books on plagiarism; and teaching students about research (including how to cite information from the Internet, and how to judge the quality of web material). Related information and policies from other American universities are included within the links.

The links under 'teaching students about plagiarism' are particularly useful. They include further links that may be printed out for students with information on what plagiarism is, what paraphrasing is and how to do it correctly, and how to reference properly. The information contained within the links on this site would be useful to both secondary educators and their students.

Valle Verde Library Combating Cybercheating: Resources for Teachers
http://www.epcc.edu/vvlib/cheat.htm

This web page is part of the El Paso Community College Valle Verde Library, and can be viewed either in English or Spanish. It includes various links to relevant web pages under specific headings. The different sections include: the evolution of the cybercheating phenomenon (includes links online articles); defining plagiarism and proper citing – guides for students; examples of plagiarised versus non-plagiarised sentences and paragraphs; combating cybercheating – strategies for teachers; and plagiarism detection services (a useful review of mostly commercial sites, some with free trials).

The information within this site has been created for the use of both students and educators. The content is relevant to both secondary and tertiary education sectors.

Noreen Reale Falcone Library: Electronic Plagiarism Seminar
http://www.lemoyne.edu/library/plagiarism/index.htm

This site was created by the librarian at Noreen Reale Falcone Library, Le Moyne College, New York, from the content of a presentation she gave on plagiarism.

The information is comprehensive, unambiguous and informative. Section headings include: plagiarism in the news; definitions; copyright; bibliography (general); bibliography (scientific misconduct); preventing plagiarism; detecting plagiarism; guides for educators; guides for students; searching tips; free papers; papers for sale; plagiarism detection sites; policies and procedures; and how to cite this web page.

Each of the links provides extensive relevant information for both students and educators. However, information in the links under the headings 'Searching for Term Paper sites' and 'Free papers', although meant to be educational, could quite easily assist those with ulterior motives to track down a particular essay they are looking for to claim as their own. This problem should, however, be managed with prior education of plagiarism and its consequences before advising students to visit this site.

Conclusion

Plagiarism is the act of using another individual's material as if it were your own, without providing appropriate acknowledgement. Plagiarism is not a new problem; however, the growth of technology use in education has led to new educational opportunities, and has created the chance for individuals to plagiarise easily from online sources. New industries have arisen in the form of paper dealers and software-detection manufacturers in response to the availability of online material that may be plagiarised.

Educators today need to be armed with a toolbox of skills to detect and prevent online plagiarism within their students' work. Educators also need to be aware of the opportunities the Internet provides for their students to plagiarise, including recognition of paper dealers and how they operate. We note that discussion groups provide anonymous professional development opportunities for both educators and students to enhance their understanding of plagiarism issues.

Prevention is the most important component in reducing incidences of plagiarism.

Plagiarism as a concept is not automatically understood by students. Students need to be able to discuss and explore notions of authorship in a safe environment. It is important for teachers to educate students on how to cite, paraphrase and to review material critically. Good assessment design that is both creative and varied is a highly recommended prevention strategy. Proving plagiarism is not a black-and-white issue and teachers need to be familiar with relevant online content to be able to follow up on student work that seems to be out of place or character.

The Internet provides many good online resources to help educators develop prevention strategies to use with their students, including lesson plans, citation guides, and sites that address issues and solutions regarding online plagiarism. We have referred to these as 'one-stop shops'.

References

Boehlert, E. (2003). *The forbidden truth about Jayson Blair*. Retrieved 13 June 2004, from: <http://www.salon.com/news/feature/2003/05/15/nytimes/index_np.html>

Brown, V.J., & Howell, M.E. (2001). The efficacy of policy statements on plagiarism: Do they change students' views? *Research in Higher Education, 42*(1), 103–118.

Centre for the Study of Higher Education (CSHE) (2002). *Plagiarism detection software: How useful is it?* Retrieved 13 June 2004, from University of Melbourne: <http://www.cshe.unimelb.edu.au/assessinglearning/03/Plag2.html>

DeVoss, D., & Rosati, A.C. (2002). 'It wasn't me, was it?' Plagiarism and the web. *Computers and Composition, 19*(2), 191–203.

Foster, A.L. (2002). Plagiarism-detection tool creates legal quandary. *The Chronicle of Higher Education, 48(36), A.37–A38.*

Galus, P. (2002). Detecting and preventing plagiarism. *Science-Teacher, 69*(8), 35–37.

Gajadhar, J. & Brooke, S. (2001). Online plagiarism: Implications for educators. *Communications Journal of New Zealand, 2*(1) 43–54.

Gibelman, M., Gelman, S.R. & Fast, J. (1999). The downside of cyberspace: Cheating made easy. *Journal of Social Work Education, Fall 1999*, 367–76.

Hawkes, M. (2000). Structuring computer-mediated communication for collaborative teacher development. *Journal of Research and Development in Education, 33*(4), 268–77.

Landau, J.D., Druen, P.B. & Arcuri, J.A. (2002). Methods for helping students avoid plagiarism. *Teaching of Psychology, 29*(2), 112–115.

McCabe, D. (2001). Cheating: Why students do it and how we can help them stop. *American Educator, 25(*4), 38-43. Retrieved 8 June, 2004, from: <http://www.aft.org/pubs-reports/american_educator/winter2001/Cheating.html>

McClain, R. & Khan, J. (2001, June 29). Cite your sites! Exploring the Internet's role in academic plagiarism. *Daily Lesson Plan*. Retrieved June 13, 2004, from: <http://www.nytimes.com/learning/teachers/lessons/20010629friday.html>

McMurtry, K. (2001). E-cheating: Combating a 21st century challenge. *T.H.E. Journal, 29*(4), 36–8.

Ministry of Education. (2002). ICT learning experience: A plague on plagiarism. *ICT Professional Development Clusters Material*. Retrieved 13 June 2004 from: <http://www.tki.org.nz/r/ict/ictpd/plague_on_plagiarism_e.php>

NMC Library (2002). *Plagiarism resources and services: A guide for faculty, Northwestern Michigan College*. Retrieved 13 June 2004 from: <http://www.nmc.edu/library/faculty/plagiarism/#definition>

Pearson, G. (2003). *Electronic Plagiarism Seminar*. Retrieved 13 June 2004 from: <http://www.lemoyne.edu/library/plagiarism/index.htm>

Petress, K.C. (2003). Academic dishonesty: A plague on our profession. *Education, 123*(3), 624–7.

Price, M. (2002). Beyond 'gotcha!': Situating plagiarism in policy and pedagogy. *College Composition and Communication, 54*(1), 88–115.

Royce, J. (2003). Has turnitin.com got it all wrapped up? *Teacher Librarian, 30*(4), 26–30.

Scanlon, P.M. (2003). Student online plagiarism: How do we respond? *College Teaching, 51*(4), 161–5.

Scanlon, P.M. & Neumann, D.R. (2002). Internet plagiarism among college students. *Journal of College Student Development, 43*(3), 374–85.

Underwood, J. & Szabo, A. (2003). Academic offences and e-learning: Individual propensities in cheating. *British Journal of Educational Technology, 34*(4), 467–77.

University of Otago Library (2002). *Checklist of criteria for evaluating websites.* Retrieved 13 June 2004 from: <http://www.library.otago.ac.nz/pdf/wc_checklist_evaluation.pdf>

University of Otago Library (2003). *E-reference.* Retrieved 13 June 2004 from: <http//www. library. otago.ac.nz/resources/vitref.html>

Vonderwell, S. (2003). An examination of asynchronous communication experiences and perspectives of students in an online course: a case study. *The Internet and Higher Education, 6(1), 77–90.*

Walker, J. & Taylor, T. (2002). *Basic CGOS Style.* Retrieved 13 June 2004, from: <http://www.columbia. edu/cu/cup/cgos/idx_basic.html>

Warlick, D. (2000). *Slate citation machine.* Retrieved 13 June 2004 from: <http://www.landmark-project.com/citation_machine/index.php>

Young, J. R. (2001, July 6). The cat-and-mouse game of plagiarism detection. *The Chronicle of Higher Education.* Retrieved 13June 2004, from: <http://www.chronicle.com/free/v47/i43/43a02601.htm>

Young, J.R. (2002, March 12). Anti-plagiarism experts raise questions about services with links to sites selling papers. *The Chronicle of Higher Education.* Retrieved 13 June 2004, from: <http://www. chronicle.com/free/2002/03/2002031201t.htm>

About the Contributors

Dr Bill Anderson is a senior lecturer in the Department of Learning and Teaching at Massey University. He works in the areas of distance education and teacher education, with specific focus on the use of information and communication technology (ICT) in education. His current research interests lie in the area of online education. He has a long-term interest in the use of educational technologies to enhance learning and teaching. Bill has been involved in the development of software for the creation of CD-Rom resource delivery for extramural students and, most recently, investigation of the potential of electronic portfolio use by students, to contribute to the goals of programmes of study as well as those of individual papers. Bill's teaching complements his research and publication in the field of online and distance education, where he has a particular interest in online interaction. He is joint editor of the *Journal of Distance Learning*, co-editor of the *Handbook of Distance Education* with Dr Michael Moore of Pennsylvania State University and has recently published in *Open Learning* and the *International Review of Research in Open and Distance Learning*. Bill's email address is: w.g.anderson@massey.ac.nz

Dr Mark Brown is a senior lecturer in the Department of Learning and Teaching at Massey University. His research interests include e-Learning, the use of ICT in education, and the study of new innovations in teaching. Mark is a trained primary teacher and has taught courses in the general field of educational technology for over fifteen years and is an Apple Distinguished Educator. Mark has a keen interest in policy analysis and has adopted a critical view of the forces shaping the ICT-related school reform movement. He is involved in pre-service teacher education from early years to secondary and coordinates two specialised postgraduate qualifications in the study of educational technologies. He is a member of Massey University's Online Learning Monitoring Group, which has responsibility for the implementation of e-Learning across the university. Mark's email address is: m.e.brown@massey.ac.nz

Liz Butterfield MNZM is the director and founder of the Internet Safety Group (ISG). She has managed the ISG's NetSafe Programme from its establishment in 1998; a programme that is today a well respected international cybersafety organisation. She has developed the NetSafe website, managed the *Hector's World* initiative aimed at three to ten-year-olds, and co-authored the *NetSafe Kit for Schools* (best practice for New Zealand schools). Liz has given presentations on cybersafety and the NetSafe programme throughout New Zealand, as well as in the United States, Vanuatu and Australia, and has written numerous articles for New Zealand, British and American publications and professional journals. She was Chair of the NetSafe II International Conference in Auckland in 2003 and is on the organising committee of a cybersafety and security conference to be held at the University of Oxford in 2005. Liz's email address is: lizb@netsafe.org.nz

Nola Campbell is a senior lecturer in ICT at the School of Education, University of Waikato. As President of the Distance Education Association of New Zealand (DEANZ), Nola has worked to spread the e-Learning message to educators around New Zealand and internationally. She has experienced e-Learning both as a student and as a teacher educator working with teachers from early childhood though to tertiary education. Her work and research in online teaching and learning reflects her interest in issues of equity, empowering people to overcome their technophobia and in working with people who are new to e-Learning. Currently Nola is Director of the Flexible Learning Leaders in New Zealand Project that is providing a programme for professional development and leadership for tertiary staff working in the e-Learning education community. Nola's email address is: ngc@waikato.ac.nz

Merrin Crooks-Simpson has worked in various ICT roles within both the private and public sector and is currently completing a Master of Arts in Education at the University of Otago, exploring women's experiences in tertiary computing education: issues regarding gender and neo-liberal reform. Her research interests include women and computing, gender issues, and ICT use within education. Merrin's email address is: mercat@clear.net.nz

Shannon Curran is a qualified secondary school teacher and is currently employed by the Student Computing Services division of the University of Otago as a tutor-supervisor working with tertiary students. She is concurrently undertaking her PhD in Education, also through the University of Otago. Her research interests include gender equity in sport (specifically rugby union in New Zealand), and the relationship between technology developments and increased rates of childhood obesity. Shannon's email address is: shannon. curran@stonebow.otago.ac.nz

Dr Garry Falloon is presently on contract to the Ministry of Education from the Faculty of Education at the University of Auckland. He is managing the Ministry's Digital Opportunities projects which aim to combine the knowledge and resources of key ICT businesses with innovative, leading edge school-based projects examining how technologies can be used to support student learning. During 2003–04, he completed doctoral research in the nature of teacher and student work practices in a 'digital classroom' environment, and hopes to be able to continue with research interests in this rapidly developing field in the future. Garry's email address is: vicandgarry@xtra.co.nz

Dr Vince Ham is director (research) at Ultralab South. Vince's research interests are in ICT in education, educational programme evaluation research methods, and teacher professional development. He has led the research teams for several Ministry of Education teacher professional development contracts, including the evaluations of school clusters' ICT professional development (ICTPD) programmes and the first two years of the *Te Kete Ipurangi* (TKI) website. He is currently a member of the editorial boards of one national and two international journals, and has served on several Ministerial and New Zealand Qualifications Authority (NZQA) advisory groups related to ICT.

Stephen Hovell is the principal of Pamapuria School, where he has spent most of his teaching career. He became interested in computers while teaching in very remote schools, before the introduction of the Internet, using computers to help students write booklets of simple rural experiences for their language programme. He completed his Master of Education at the University of Otago where he undertook studies in ICT in an educational context. His special area of interest is Virtual Field Trips, which formed the topic of his M.Ed. research. Currently he is investigating the nature of primary school science and how teachers can help children refine their understanding of scientific concepts. Stephen's email address is: srh@xtra.co.nz

Dr Kwok-Wing Lai is an associate professor in the Faculty of Education, University of Otago. Wing has a keen interest in studying and researching e-Learning and the use of ICT to facilitate the development of online communities of practice for teachers. He is the founding editor of *Computers in New Zealand Schools*. Wing's email address is: wing.lai@stonebow.otago.ac.nz

Dr Judy Parr is a co-director of the Research Centre for Interventions in Teaching and Learning at the University of Auckland. Her research and teaching interests are in developmental and educational psychology, particularly literacy with emphasis on writing and the interface between literacy and technology (although this latter interest broadens to include other aspects of technology in educational settings). Recent articles appear in *Language Arts*, *The Reading Teacher* and the *International Journal of Educational Research*. Judy co-wrote *Using evidence in teaching practice* published in 2004. She has been the principal researcher in numerous projects in the last few years including: a review of computer-assisted instruction in literacy and numeracy; effective practice in use of readymade literacy materials in classrooms; assessment tools for teaching and learning in writing; evaluating Digital Opportunities; FarNet; and an evaluation of the national literacy leadership initiative. Judy's email address is: jm.parr@auckland.ac.nz

Dr Keryn Pratt has been working in the Faculty of Education, University of Otago, since 2000, first as a research fellow and more recently as a lecturer. She teaches a number of web-based papers in ICT in education. Keryn's experience and interest in ICT and education includes children's skills and usage and attitudes towards ICT, professional development of teachers, ethical, health and safety issues associated with the use of ICT and the effect ICT can have for teaching and learning. She has worked collaboratively on a number of large-scale research projects evaluating ICT use by teachers and students in New Zealand. Keryn's email address is: keryn.pratt@ stonebow.otago.ac.nz

Dr Ken Ryba is the co-ordinator of the College of Education at Massey University's Albany Campus. Ken recently established a new postgraduate programme for training educational psychologists and he is well known for his research and leadership in the special education field. Ken was one of the earliest researchers to apply ICT as a learning tool for students with special needs and has been involved in a number of ICT projects in schools. He has published extensively in the field of educational computing. Ken's email address is: K.A.Ryba@massey.ac.nz

Dr Linda Selby is the dean, Postgraduate Studies and Research in the Faculty of Education at the University of Auckland. She has a background in primary and secondary teaching and has taught in New Zealand, Australia and Canada. Her research interests include the use of ICT in educational settings, teaching and learning with the Internet, equity issues and computing, ICT and special education, and teacher professional development.

She is the author of several book chapters and articles, a member of the New Zealand Ministry of Education ICT Research Advisory Group and the editor of *Computers in New Zealand Schools*. Linda's email address is: l.selby@ace.ac.nz

Lorrae Ward is a part-time lecturer and PhD student at the University of Auckland. Her research interests are in the areas of school change, leadership and professional learning, with a particular interest in the implementation of ICT initiatives. She has worked on a number of evaluation projects including FarNet and Innovations Funding. Lorrae's email address is: l.ward@auckland.ac.nz

Russell Yates is a senior lecturer in Professional Studies in Education at the University of Waikato. His background includes experience as a primary school teacher and principal and a lecturer in professional education. Russell's experience in online education is in the Mixed Media Programme that he coordinates. The programme, which began in 1997, provides online opportunities for students who cannot attend campus classes to become primary school teachers. He has a particular interest in establishing and maintaining support systems for students and the schools they work in. Russell has presented and researched work in online student support. His email address is: ryates@waikato.ac.nz

Index

Facilitators 52, 53, 56, 57, 61, 62, 63, 64; qualities of 65–6; duties 99, 100; full-time/part-time 66–7

FarNet 95, 125–6, 127–33

Firewalls 17, 186

FitNet project 97

Gender 30–31, 68, 76, 77, 80, 111, 112, 115, 116–18, 119, 120, 121, 164, 166

GenXP project 95

Globalisation 28, 29, 34

Hammer, Law of the 14

Health risks of ICT use 16–17, 85

Hidden curriculum 24, 32

ICT 9–20, 24, 29, 42, 45 see also Access to ICT; Health risks of ICT; Integration of ICT: educational reform 23, 27; funding 9–10; implementation 75

ICT coordinators 13, 31, 66, 75–90, 126, 157–8; curriculum support 85, 87–8; duties 77, 78, 82–8; funding 90; gender issues 31, 76, 77, 80; integration of ICT 81, 86, 90; ICT use 81, 86; leadership role 88, 90; professional development 81, 85–6, 88, 89–90; technical support role 79, 82, 87, 89, 90; time issues 79, 88–9, 90

ICT policies 28–9, 32, 34, 54, 93; school level 83, 84, 86–7

ICT professional development (ICTPD) 9, 10, 18–20, 32, 47, 131; ICT coordinators 81, 85–6, 88, 89–90

ICT use 9, 10, 13, 14, 25, 26, 33, 73–4, 125, 148, 157–8, 169; ICT coordinators 81, 86; classroom strategies 61; gender issues 30–31; leadership 12–13; management 13; pedagogical beliefs 12, 15; pedagogical benefits 12, 14, 25–7, 29; pedagogical rationales 25, 29, 61–2; pedagogy 13, 14, 15; school culture 12; teacher confidence 67–9

ICTPD clusters 32, 51–69, 96, 101, 104, 125; action plans 61; cluster size 56; facilitator qualities 65–6; funding 67; locality 57; modes of delivery 62–4; origins of scheme 53–5; teacher confidence 67–9; time issues 57–60, 63, 68

In-service teacher education 51, 52; internet-based 54

Information and communication technology see ICT

Information highway 24, 26

Information/knowledge dichotomy 18

Innovation 129

Instant messaging 184

Integration of ICT 10, 12, 13, 14–15, 18, 27, 73, 74, 75, 78, 101, 102, 131, 146, 152; cybersafety 190; ICT coordinators 81, 86, 90; pedagogical beliefs 12, 15; school culture 12

Integration of technology in the school curriculum see Integration of ICT

Internet 112, 130, 131, 180, 181, 182, 183, 189 see also Computer-mediated communication; Cybersafety; Cyberspace; ICT; Searching for information

Internet access 9, 10, 24, 157–8, 169, 181, 193

Internet connectivity 9, 10, 73

Internet Relay Chat (IRC) 188

Internet Safety Group 181, 182, 184, 185, 190–93, 194, 195; NetSafe programme 190–92, 193, 195; safety button 193

Knowledge economy 29, 33, 34

Laptop computers 12, 35, 96, 192; health risks 16

Lead schools 55, 57

Lead teachers 60, 64, 66, 98, 126

Leadership 12–13, 64, 65, 67, 74–5, 76; ICT coordinators 88, 90; principal's role 75, 76, 90; shared 75

LEARNZ virtual field trips 135, 136, 137, 140, 141, 142–52; access to 150–51; ambassador programme 145, 149, 151, 152; audioconferencing 145, 147, 150, 151, 152; authentic contexts 145, 147; collaborative learning 149–50; curriculum links 144, 146, 147; evaluation 146–8; pedagogy 142, 143; problem solving 148, 149; professional development 143; scaffolding 149; time management 151; web boards 145

Librarians 41, 48, 169, 192

Libraries 41, 48, 97

Linking education with Antarctic research in New Zealand see LEARNZ

Listservs 126, 131, 141

Maine Learning Technology Imperative (MLTI) 35

Maori curriculum group 128

Maori curriculum leaders 128

Maori curriculum site 128

Maori medium schools 101

Maori teachers 128, 132

Mentoring 19, 44, 62, 63, 132

Metaphors of learning 28, 102–3